Measurement and Use of Shear Wave Velocity

for Evaluating Dynamic Soil Properties

Proceedings of a session sponsored by
the Geotechnical Engineering Division of
the American Society of Civil Engineers
in conjunction with the ASCE Convention
in Denver, Colorado

May 1, 1985

Edited by Richard D. Woods

Published by the
American Society of Civil Engineers
345 East 47th Street
New York, New York 10017-2398

The Society is not responsible for any statements made or opinions expressed in its publications.

Library of Congress Catalog Card No.: 85-70486
ISBN 0-87262-456-0
Manufactured in the United States of America.

FOREWORD

Geotechnical engineers have recognized in the past decade that shear wave velocity is a basic soil property and have begun to use it to characterize sites for many uses. Most notable has been the realization that the so called "dynamic modulus" is simply the low strain value of elastic modulus and this can be used in many "static" as well as "dynamic" applications. Because of its basic nature, it has also been recognized that shear wave velocity is an excellent diagnostic tool which can be used to evaluate the results of soil modification techniques.

Three of the papers submitted to this session deal with the measurement of shear wave velocity; one through its relationship with Rayleigh Waves, another by coupling with cone penetration sounding, and the third by combined field and lab measurements. All represent significant advances in the art of shear wave velocity measurements but for different reasons. The papers by Stokoe and Nazarian, and Stoll begin to take advantage of the long known connection between time and frequency domains and the powerful relationships provided by Fourier Duals. Geotechnical engineers are finally beginning to recognize and take advantage of some of the tools which the geophysicist has been using for several decades. The paper by Robertson et al also provides an important marriage between well known in-situ exploration techniques, cone penetration tests and down-hole seismic tests. The combining of these two should result in data which is greater than the sum of the two methods separately.

Two of the papers deal with an application of shear wave velocity in studing one of the most important problems of geotechnical engineering, liquefaction, either in the identification of the potential for liquefaction as demonstrated by Stokoe and Nazarian or in the evaluation of the stability of an earth embankment subject to earthquake shaking as described by Charlie et al.

As a compliment to the other four papers in this session, the one by Lew and Campbell focuses our attention on an empirical correlation between shear wave velocity and depth of overburden. Information of this type from firms or agencies who often perform shear wave velocity tests is very useful in observing the overall impact of a specific soil property measurement on the daily work of our profession.

Each of the papers included in the Proceedings received two positive peer reviews. All papers are eligible for discussion in the Journal of Geotechnical Engineering and all papers are eligible for ASCE awards.

The Soil Dynamics Committee of the Geotechnical Engineering Division is pleased to sponsor this session on "Measurement and Use of Shear Wave Velocity For Evaluating Dynamic Soil Properties". This committee has sponsored many sessions at past ASCE conventions which have aided in disseminating knowledge about many aspects of soil dynamics and this session continues that tradition. Soil dynamics has matured rapidly over the past two decades attaining the status of an essential part of more and more geotechnical projects. The soil dynamics committee is pleased to have been a part in this development.

Richard D. Woods
Editor

FOREWORD

CONTENTS

USE OF RAYLEIGH WAVES IN LIQUEFACTION STUDIES

Kenneth H. Stokoe, II[1], M. ASCE and Soheil Nazarian[2], S.M. ASCE

ABSTRACT

The Spectral-Analysis-of-Surface-Waves method is a new seismic method for efficient determination of shear wave velocity in situ. Vertical impacts applied to the ground surface are used to generate Rayleigh waves with different frequencies. Propagation of the waves is monitored with surface receivers located known distances apart. By employing waveform analysis techniques, the variation of Rayleigh wave velocity with frequency is calculated. With inversion, shear wave velocity, shear modulus and layering of the site are determined. Several case studies where liquefaction occurred are presented. In each case, the shear wave velocites of the liquefiable layer were less than 450 fps.

INTRODUCTION

Shear wave velocity and its variation with shearing strain are essential in characterizing nonlinear soil behavior during earthquake shaking. One of the key elements in characterizing this behavior is measurement of shear wave velocity in the field at small strains, shearing strains less than 0.001 percent. Seismic techniques such as the crosshole and downhole tests are presently employed for this task. However, these in situ tests are under-utilized because of economic, time and personnel considerations.

A new seismic method for in situ measurement of low-amplitude shear wave velocities of soil deposits and thicknesses of soil layers is presented herein. This method is called the Spectral-Analysis-of-Surface-Waves (SASW) method. The SASW method is based upon generation and measurement of surface waves, Rayleigh waves. The method is fast, economical, nondestructive and requires no boreholes. In addition, the method has the potential of nearly full automation; hence, the need for specialized personnel to perform the test and analyze the data is minimized.

The theoretical background, digital signal processing, field testing technique and data reduction procedures associated with SASW testing are discussed herein after which several case studies are presented. The case studies all involve areas where liquefaction occurred in southern California in either the 1979 Imperial Valley Earthquake or the 1981 Westmorland Earthquake.

---

[1] Professor, Civil Engr. Dept., Univ. of Texas, Austin, Texas 78712
[2] Research Associate, Univ. of Texas, Austin, Texas 78712

THEORETICAL BACKGROUND

Elastic Wave Propagation

A vertical impulse applied to the surface of an elastic half-space generates two types of waves; body waves and surface waves. Body waves consist of compression and shear waves (P- and S-waves, respectively) which propagate along hemispherical wavefronts into the half-space. On the other hand, surface waves (Rayleigh waves, R-waves) only travel near the surface and only along cylindrical wavefronts. Body and surface waves also differ in the amount of geometrical damping associated with each wave. Near the surface, body waves attenuate proportional to $1/r^2$ (where r is the distance from source), but Rayleigh waves attenuate more slowly, proportional to $1/\sqrt{r}$. The SASW method makes use of the lower attenuation and cylindrical wavefront characteristics of Rayleigh waves. In addition, the fact that the majority of the energy from a vertical impact goes into Rayleigh wave energy further benefits the method.

In a homogeneous elastic half-space, R-wave velocity is constant and independent of frequency. In this case, the frequency of excitation, f, and R-wave velocity, V_R, are related by:

$$V_R = f \bullet L_R \tag{1}$$

where, L_R is wavelength. The components of R-wave particle motion are distributed nonuniformly with depth. At a depth approximately equal to one wavelength, particle motions are quite small. Thus it can be assumed that material to a depth of approximately one wavelength is the material predominantly sampled. From Eq. 1, frequency and wavelength are inversely proportional. As such, high frequencies are associated with short wavelengths and vice versa. In other words, high frequencies sample near-surface material while lower frequencies sample deeper materials.

If the stiffness of the medium varies with depth, R-wave velocity will become dependent on frequency, and different frequencies will propagate with different velocities. In an oversimplified fashion, for each frequency an average velocity of the material down to a depth of approximately one wavelength is measured. This velocity is termed phase velocity or apparent R-wave velocity. The fact that different frequencies propagate with different velocities is known as the dispersive characteristic of surface waves. This dispersive characteristic is a key element in SASW testing because it allows different materials to be sampled by using different wavelengths.

A plot of phase velocity versus frequency or wavelength is called a dispersion curve. Such a curve can be considered the raw data in SASW testing. Once obtained, the curve must be inverted. Inversion of a dispersion curve is the process of obtaining a shear wave velocity profile from a dispersion curve. Inversion can be viewed as a process of identifying the stiffness and thickness of each layer from the dispersion curve. The simplest method of inversion is to assume that the shear wave velocity is approximately equal to about 110 percent of the phase velocity and the effective sampling depth for each wavelength is equal to a fraction of that wavelength. Ballard (1964) and Ballard and Casagrande (1967) recommended that 1/2 of the wavelength is representative of the effective depth of sampling. Gazetas (1982), based on theoretical studies, and Heisey, et al (1982), based upon several experiments, recommended 1/3 of the wavelength as the effective depth of sampling.

Inevitably use of this simple inversion method results in some errors due to the nature of the assumptions. Existence of a layer with relatively high or low velocity near the surface causes a shift in the measured velocities of the underlying layers towards higher or lower velocities in the dispersion curve. If this contrast in velocities is relatively small, then the simple inversion method may work reasonably well. However, use of the simple method normally results in shear wave velocity profiles which are doubtful. As such, a refined inversion method based upon the Haskell (1953) and Thomson (1950) matrix for multi-layered media with some modifications has been developed. In this method the medium is divided into N layers as shown in Fig. 1. Due to an impact on the ground surface, seismic waves propagate in the medium. Theoretically, for a harmonic wave, the wave equation for layer n can be written as:

$$\Phi_n = U_{pn} \exp [ir_n(z - z_{n-1})] + D_{pn} \exp [-ir_n(z - z_{n-1})] \quad (2)$$

$$\Psi_n = U_{sn} \exp [is_n(z - z_{n-1})] + D_{sn} \exp [-is_n(z - z_{n-1})] \quad (3)$$

where,

Φ_n = a potential corresponding to compression waves,

Ψ_n = a potential corresponding to shear waves,

r_n = $\sqrt{k^2 - k^2_{pn}}$,

s_n = $\sqrt{k^2 - k^2_{sn}}$,

k = $2\pi f/V_{ph}$, wave number for surface waves,

k_{pn} = $2\pi f/V_{pn}$, wave number for compression waves,

k_{sn} = $2\pi f/V_{sn}$, wave number for shear waves, and

f = frequency of excitation.

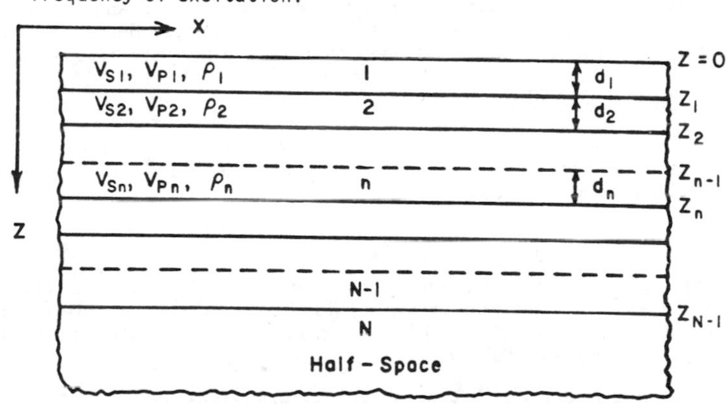

Fig. 1 - Idealized Model of a Heterogeneous Medium.

Parameters U_{pn} and D_{pn} correspond to magnitudes of upgoing and downgoing P-waves in the nth layer, respectively. Similarly, U_{sn} and D_{sn} are magnitudes of upgoing and downgoing S-waves in the nth layer, respectively. These four parameters are unknown and should be determined. To determine them, the displacements and stresses of each soil element are determined and related to potentials Φ_n and Ψ_n by utilizing elastic wave propagation theory.

If the condition of plane waves is assumed, the normal and shear stresses acting on a horizontal plane at each point in the medium as well as vertical and horizontal components of displacement can be calculated. At the boundary of each layer, the stresses and displacements must be continuous. In other words, the stress and displacement components at the top interface of a layer should be equal to those at the bottom interface of the adjacent, overlying layer. Continuity of stresses and displacements at the boundaries yields four boundary conditions for each layer. These four boundary conditions are used to determine the four parameters U_{pn}, D_{pn}, U_{sn} and D_{sn} for each layer. Additional boundary conditions are: 1) the stress components at the ground surface should be equal to zero; and, 2) for the last layer, which is considered to extend to infinity in the vertical direction, waves cannot propagate upward (radiation condition).

The solution of these simultaneous differential equations yields a relationship between the frequency of excitation and phase velocity called a dispersion equation. The dispersion equation is then used to generate a theoretical dispersion curve. Once the theoretical and measured (experimental) dispersion curves match, the shear wave velocity profile is determined uniquely. This process is a key element in SASW testing and has been missing from previously used Rayleigh wave methods.

DIGITAL SIGNAL ANALYSIS

In recent years with the development of digital data acquisition systems and the development of the Fast Fourier Transform algorithm (FFT), collecting and processing of data has been revolutionized. Use of the FFT and spectral analysis can substantially decrease testing times by replacing numerous steady-state experiments with a single transient event. Such analyses are also key elements in SASW testing. Hence, the different terms and definitions used in SASW testing are briefly discussed.

Fast Fourier Transform

Transformation of data from the time domain to the frequency domain is based on the Fourier transform (Hewlett-Packard, 1981). The Fourier transform is defined as:

$$S_x(f) = \int_{-\infty}^{\infty} x(t) \exp(-2\pi i f t) dt \tag{4}$$

where, $S_x(f)$ is the frequency domain representation of the function x(t), and $i = \sqrt{-1}$. If the function x(t) is digitized, the Fourier transform of this digitized function, which is called the Discrete Finite Transform (DFT), can be presented by:

$$S_x(m\Delta f) \simeq \Delta t \sum_{n=0}^{N-1} x(n\Delta t) \exp[-2i\pi(m\Delta f)(n\Delta t)] \tag{5}$$

where:

$S_x(m\Delta f)$ = the digitized representation of the Fourier transform,
N = number of digitized points,
Δt = time interval between digitized points,
m = 0, ±1, ±2, ... and
Δf = frequency interval between digitized points.

The time and frequency intervals are related by:

$$\Delta f = 1/(N\Delta t) \quad (6)$$

The Fast Fourier Transform (FFT) is an algorithm for computing the DFT rapidly and efficiently and is available on many waveform analyzers. One such waveform analyzer, a Hewlett-Packard 5423A, was used in this work.

Spectral Analysis.

Many different measurements can be made once the signals are transformed into the frequency domain. The linear spectrum, $S_x(f)$, of a signal is basically the Fourier transform of the signal recorded in the time domain. The linear spectrum is used to identify the predominant frequencies and their absolute amplitudes and phases in a record. Figure 2a shows the records in the time domain of a vertical impulse recorded by two vertical geophones at distances of about 4 ft and 8 ft from the impact. The linear spectrum of Channel 1 is shown in Fig. 2b.

The auto spectrum, G_{xx}, is the product of the linear spectrum, $S_x(f)$, and its complex conjugate, $S_x^*(f)$. The magnitude of the auto spectrum is equal to the magnitude of the linear spectrum squared and can be expressed as:

$$G_{xx}(f) = S_x(f) \bullet S_x^*(f) \quad (7)$$

The auto spectrum is the Fourier transform of the auto correlation in the time domain. It can be used to find predominant periods from a time record.

If a system can be modelled as a linear system, such as soil media at low strain levels, signals from two records obtained simultaneously can be compared utilizing spectral analysis. The cross power spectrum, G_{xy}, is the product of the linear spectrum of record 1 (i.e. S_x) and the complex conjugate of the linear spectrum of record 2 (i.e. S_y^*) as such:

$$G_{xy}(f) = S_y^*(f) \bullet S_x(f) \quad (8)$$

The magnitude of the cross power spectrum is an indication of the mutual frequencies in the two records. The phase information of G_{xy} can be used to determine the phase difference between the two records using the rotating-vector principle. The cross power spectrum is a good tool in determination of the relative phase difference between two signals caused by time delays, propagation delays or varying wave paths between receivers. As such, the cross power spectrum is an important measurement in SASW testing. A cross power spectrum determined from the records in Fig. 2a is shown in Fig. 2c.

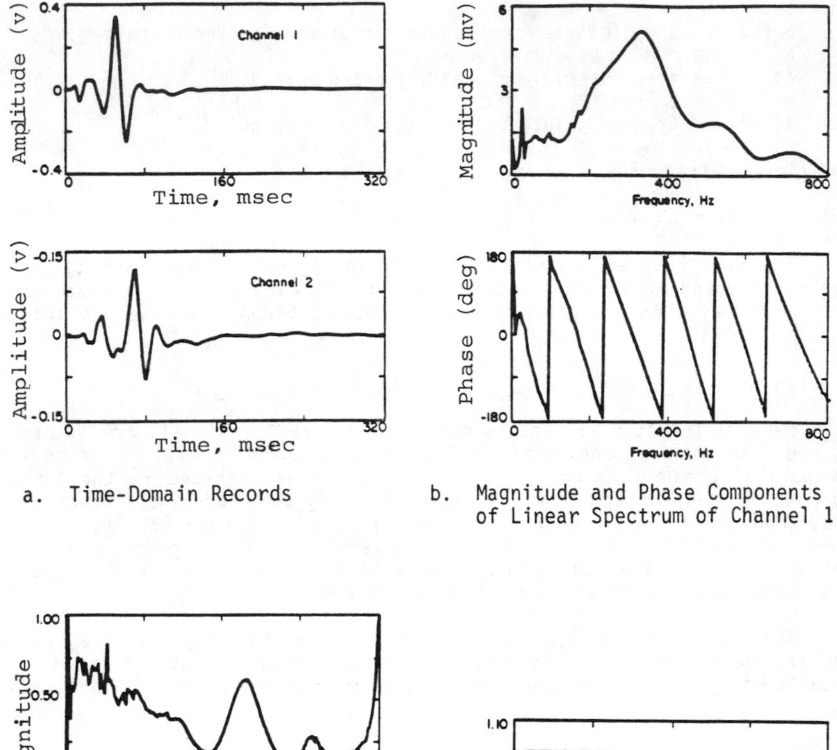

a. Time-Domain Records

b. Magnitude and Phase Components of Linear Spectrum of Channel 1

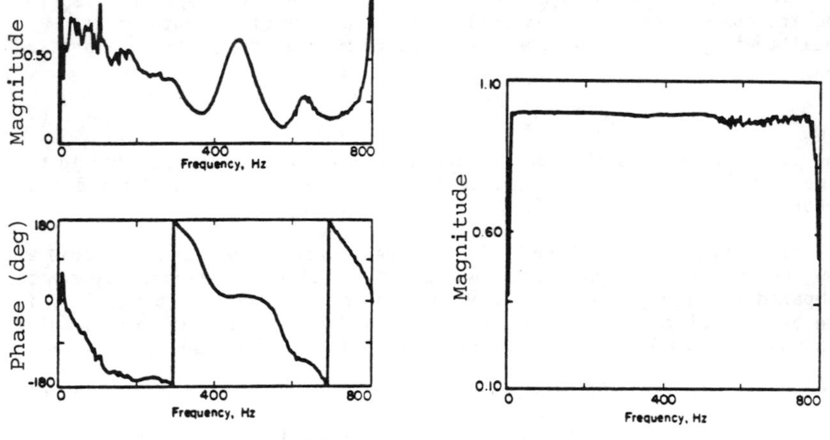

c. Magnitude and Phase Components of Cross Power Spectrum

d. Coherence Function

Fig. 2 - Typical Signal Processing in SASW Testing.

The last parameter used in SASW testing is coherence. The coherence function, $\gamma^2(f)$, is analogous to a signal-to-noise ratio. Coherence is defined as:

$$\gamma^2(f) = [G_{yx}(f) \cdot G^*_{yx}(f)]/[G_{xx}(f) \cdot G_{yy}(f)] \quad (9)$$

The signal-to-noise ratio (S/N) can be calculated from the coherence function by:

$$S/N = \gamma^2(f)/[1 - \gamma^2(f)] \quad (10)$$

The coherence is a real-valued function corresponding to the ratio of the response (output) power caused by the measured input to the total measured response power. A value of coherence equal to unity at a certain frequency corresponds to perfect correlation between the two signals and indicates that the signals are not contaminated with random background noise. Therefore, the coherence function is used to check the data in the field during data collection. The coherence function determined from the records in Fig. 2a is shown in Fig. 2d.

METHOD OF SPECTRAL-ANALYSIS-OF-SURFACE-WAVES

The Spectral-Analysis-of-Surface-Waves (SASW) method is a nondestructive method of determining the shear wave velocity profile and layer thicknesses from Rayleigh wave measurements. Data collection and in-house data reduction are based on the theoretical points and digital signal processing just discussed.

Field Testing

Two vertical velocity transducers are used as receivers in the field. A linear array is employed and an imaginary centerline is selected. The receivers are placed securely on the ground surface symmetrically about this centerline as shown in Fig. 3a. A transient impulse is transmitted to the soil by means of an appropriate hammer. The range of frequencies over which the receivers should function depends on the site being tested. To sample deep materials, 50 to 100 ft, the receiver should have a low natural frequency, typically within the range of 1 to 2 Hz. In contrast, for sampling shallow layers, the receivers should be able to respond to high frequencies in the range of 1000 Hz or more. For testing presented herein, receivers with a natural frequency of 4.5 Hz were used and sampling was done to about 40 feet.

The shape and weight of the source depend on the distance between receivers and the stiffness of the material being tested. For sampling near-surface material in which the receivers are closely spaced, a lightweight hammer capable of generating high frequencies is desirable. In this case, a small claw hammer works well. As the distance between receivers increases, heavier sources are used to generate lower frequencies. A Standard Penetration Test hammer falling directly on the soil has been used for the largest receiver spacings (over 32 feet).

Theoretically, one test is sufficient for determining the dispersion curve. However, to obtain more representative data, several tests with different receiver spacings are performed. In each test the distance

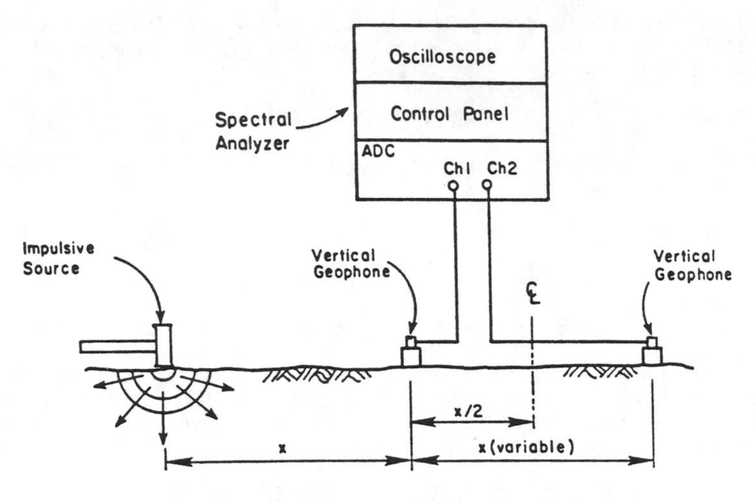

(a) General Configuration of SASW Tests

(b) Common Receivers Midpoint Geometry

Fig. 3 - Schematic of Experimental Arrangement for SASW Tests.

between the receivers is generally doubled. A typical experimental set-up is shown in Fig. 3b. The geophones are always placed symmetrically about the selected, imaginary centerline. This pattern of testing is called the common receivers midpoint (CRMP) geometry. Nazarian and Stokoe (1983) have shown that use of this set-up reduces scatter in data collection due to the fact that the distance covered in the previous tests are always included in the next tests. In addition, at each receiver spacing, two series of experiments are performed. First, the test is carried out from one direction (forward profile) and then without relocating the receivers the same test is performed with the source on the opposite side of the receivers (reverse profile). By running forward and reverse profiles and by averaging the outcomes of these two tests, the effect of any internal phase shift between receivers is minimized and the effect of dipping layers along the the distance between the receivers is averaged.

The receivers are connected to a spectrum analyzer which serves as the recording device. A spectrum analyzer is a digital oscilloscope which, by means of a microprocessor attached to it, has the ability to function in both the time and frequency domains. The analog signals generated by the receivers are digitized and saved. These signals are then fast Fourier transformed, and spectral analyses are performed on the them.

To enhance the signals, a few records are averaged. One of the advantages of the use of spectral analysis is that synchronized triggering is not required to average the records. Averaging can reduce the effect of random background noise substantially, as the average of several random events tends to be zero. Based on the writers' experience, the optimum number of averages is typically about five.

<u>In-House Data Reduction</u>

In-house data reduction consists of construction of the dispersion curve and then inversion of this curve. Spectral analysis is utilized to construct the dispersion curve. The aspects of interest are the coherence function and phase information of the cross power spectrum. A typical coherence function and phase information of the cross power spectrum are shown in Fig. 4 for a site which liquefied. This site is calld the Wildlife site hereafter. From the coherence function, the range of frequencies that should be considered in each record is selected. In this range the phase difference for each frequency is then measured. As seen in Fig. 4a, four regions with low coherence exist, and phase data in these regions was thus deleted from construction of the dispersion curve.

A phase shift equal to 360 degrees is equivalent to one period. As such, the travel time of surface waves with different frequencies can be determined by:

$$t = T \cdot (\phi/360) \qquad (12)$$

where

t = travel time associated with a certain frequency,
ϕ = phase shift for that frequency,
$T = 1/f$ = period of the frequency and,
f = the frequency.

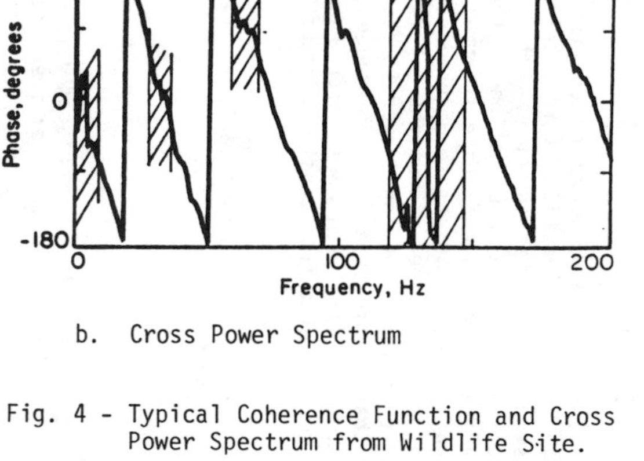

a. Coherence

b. Cross Power Spectrum

Fig. 4 - Typical Coherence Function and Cross Power Spectrum from Wildlife Site.

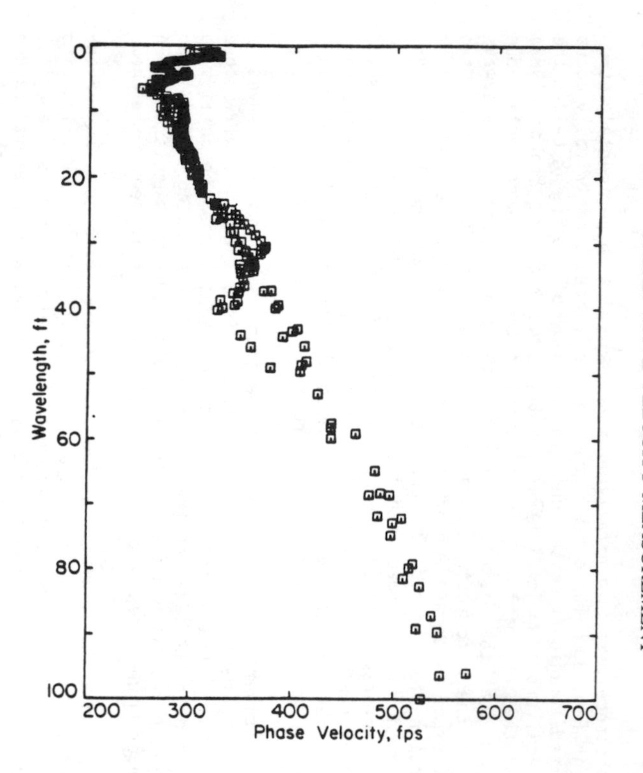

Fig. 5 - Dispersion Curve from All Receiver Spacings for Wildlife Site.

As the distance between the receivers is known,the phase velocity, V_{ph}, for a given frequency can be calculated by:

$$V_{ph} = X/t \quad (13)$$

where X is the distance between the geophones. By knowing the velocity and frequency, the wavelength, L_{ph}, is equal to:

$$L_{ph} = V_{ph}/f \quad (14)$$

By repeating this procedure for each frequency in the acceptable range of frequencies (which is selected from the coherence function) for all receiver spacings, a comprehensive and representative dispersion curve is obtained.

Heisey, et al (1982) have shown that due to limitations of recording equipment and attenuation properties of soil media any data point with a wavelength greater than three times or less than one-half of the receiver spacing should not be considered in construction of dispersion curves. However, as the distance between the receivers is doubled for each test, different sections of the dispersion curve obtained from different receiver spacings overlap and a continuous curve is determined.

A typical dispersion curve is shown in Fig. 5. This curve was determined for the Wildlife site using six different receiver spacings.

Inversion of the Dispersion Curve

Inversion is the process of determining the shear wave velocity profile from the dispersion curve. The process used in this study is to obtain iteratively a theoretical dispersion curve which matches reasonably well with the experimental dispersion curve. The theoretical dispersion curve is constructed using a modified version of the Haskell (1950) - Thomson (1953) matrix formulation which is incorporated in an interactive computer program. To initiate the process, a shear wave velocity profile is assumed. For the first trial, the shear wave velocity profile obtained by the simplified inversion method (previously discussed) is used. If the theoretical and experimental curves match, the desired profile is obtained. However, if the two curves do not match, the shear wave velocity profile is modified, and another theoretical curve is constructed. This trial and error procedure is continued until the two curves agree reasonably well.

It should be mentioned that Poisson's ratio and mass density of each layer should be estimated. However, these properties are known in a close range for different materials, and Grant and West (1965) have shown that the effect of misestimation of Poisson's ratio and mass density have little effect on the final outcome.

The experimental and theoretical dispersion curves for the last iteration for the Wildlife site are shown in Fig. 6. The curves have been expanded so that the match can be more easily seen.

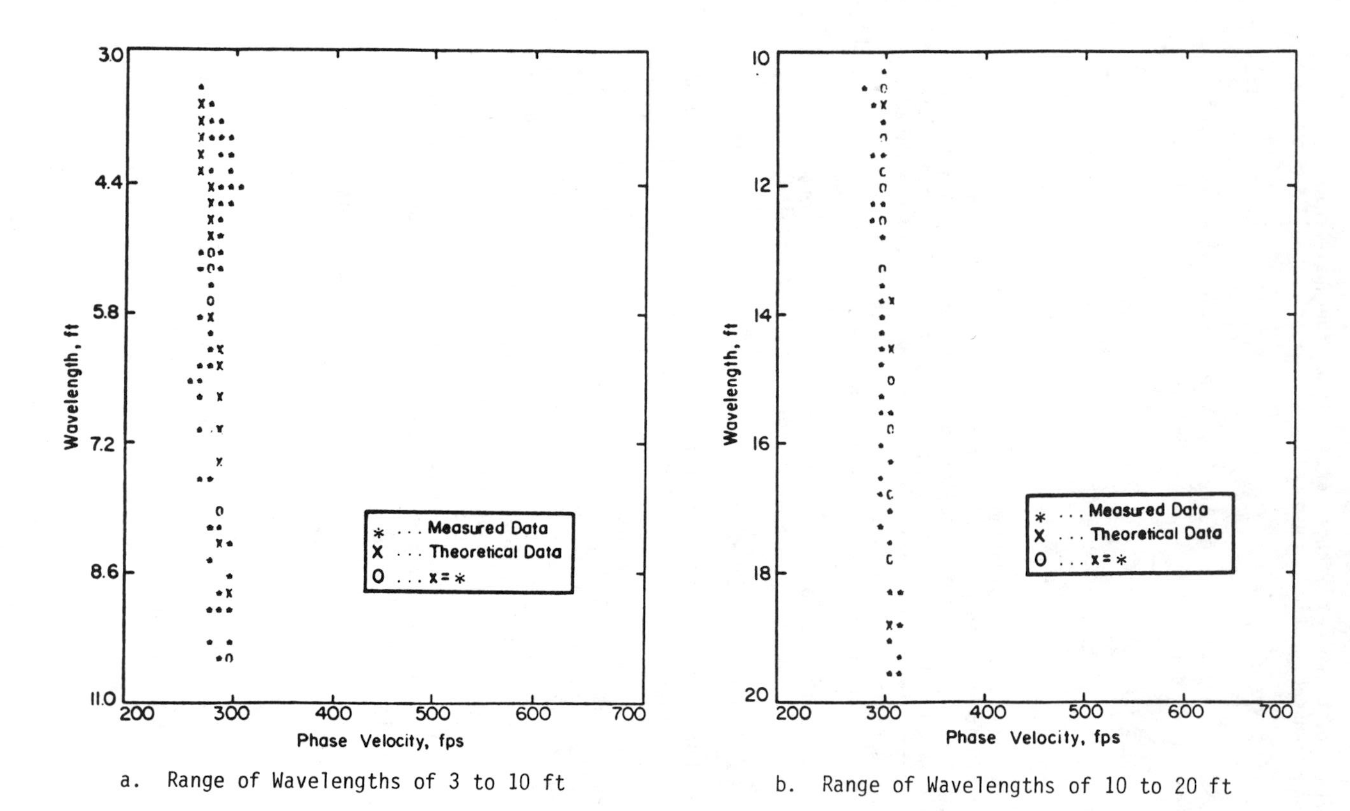

a. Range of Wavelengths of 3 to 10 ft

b. Range of Wavelengths of 10 to 20 ft

Fig. 6 - Comparison of Theoretical and Experimental Dispersion Curves after Completion of Inversion Process.

LIQUEFACTION STUDY

Recently two major earthquakes occurred in southern California; the Imperial Valley Earthquake of 1979 and the Westmorland Earthquake of 1981, with Richter magnitudes of 6.6 and 5.6, respectively. Extensive damage occurred to buildings and roads during each earthquake (Geological Survey, 1982; Youd and Wieczorek, 1981). Extensive liquefaction also occurred, the manifestation of which could be detected by numerous sand boils. Geological information of the area is presented in detail in Geological Survey (1982) and Haag et al (1985) and is not repeated herein.

As a part of a joint investigative program between the U.S. Geological Survey, The University of Texas, Rensselear Polytechnic Institute, and Woodward Clyde Consultants, an extensive field study of several sites was carried out in Imperial Valley. A plan view of the area with locations of the major faults and epicenters of the earthquakes is shown in Fig. 7. The location of the test sites is also shown in this figure. Shear wave velocity profiles were determined at all sites using the SASW method. In addition, crosshole seismic tests were performed at the Wildlife site.

The shear wave velocity profile obtained from matching the dispersion curve in Fig. 5 is shown in Fig. 8. In the inversion process, a total of 20 layers, each 1.75-ft thick, were used. The shear wave velocity profile is shown in Fig. 8 along with those obtained from the crosshole seismic tests. The two seismic profiles compare very favorably, with velocities from the two independent methods generally within about ten percent.

A composite profile of the site is shown in Fig. 9. The layering obtained from the cone penetration test (CPT) is shown in Fig. 9. Three distinct layers exist; 9 ft of clayey silt underlain by 14 ft of silty sand (which liquefied) which in turn is underlain by a clay layer extending at least to a depth of 40 ft. At this depth the test was stopped. The layering observed during the drilling operation agreed well with the layering from the CPT. However, as the stiffnesses (i.e., shear wave velocities) of the top two layers are similar, the two layers are indistinguishable from the SASW results. The two layers have different characteristics in permeability and strength resulting in different liquefaction susceptibility; but they have similar stiffnesses. Therefore, the importance of drilling boreholes or alternative tests such as the CPT should not be ignored. The boundary between the second and third layers is predicted by the SASW quite accurately as shown in Fig. 9.

Shear wave velocity profiles were determined by SASW tests at six sites which liquefied (shown as 1 through 5 and 7 in Fig. 7) in either the 1979 Imperial Valley or 1981 Westmorland earthquakes. A composite profile of the shear wave velocity profiles of the layers that liquefied at these sites is presented in Fig. 10. The soils ranged from silty sands (SM) to sandy silts (ML). As can be seen from the figure, the soils that liquefied all have shear wave velocities less than 450 fps. In fact, if only the 1981 Westmorland earthquake is considered, only sites 1 through 4 liquefied, and sands at these sites exhibit shear wave velocities less than 400 fps in the top 20 ft of depth. Further, the shear wave velocity of a silty sand at one site along Heber Road (site 6 in Fig. 7) which did

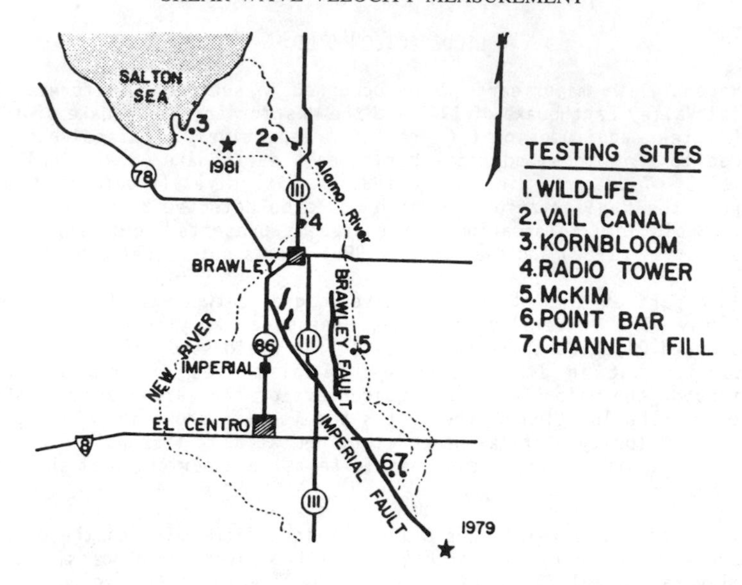

Fig. 7 - Plan View of Imperial Valley with SASW Test Sites.

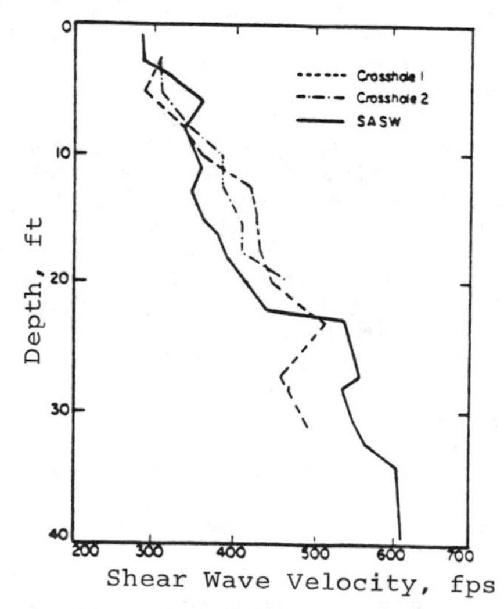

Fig. 8 - Shear Wave Velocity Profiles from SASW and Crosshole Tests at Wildlife Site.

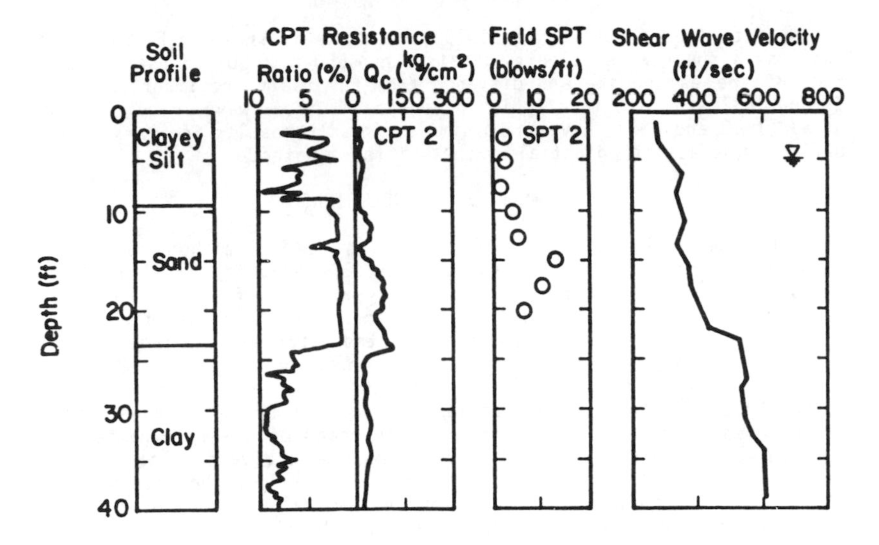

Fig. 9 - Composite Profile of Wildlife Site.

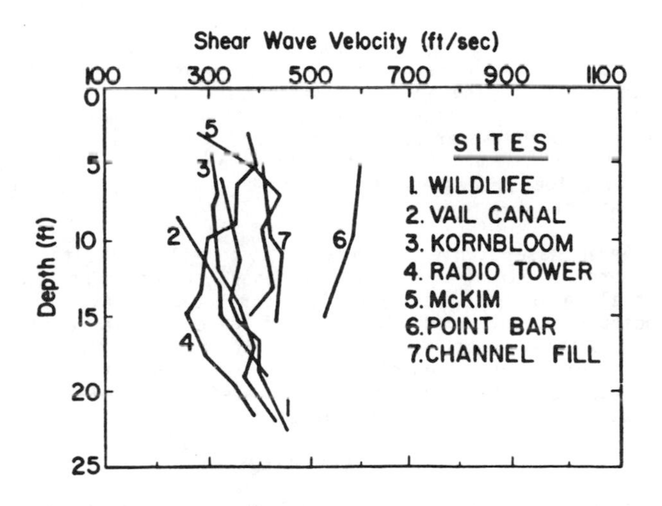

Fig. 10 - Profiles of Shear Wave Velocities of Sandy Soils which have Previously Liquefied.

not seem to liquefy during the 1979 earthquake is shown in Fig. 10. This sand has a shear wave velocity greater than 550 fps. Of course the magnitude of the earthquake and distance from the fault are important, but a strong indicator of liquefaction potential is shear wave velocity. Dobry et al (1980 and 1981) have shown similar in situ results in terms of evaluating liquefaction potential by the stiffness method.

SUMMARY AND CONCLUSIONS

The Spectral-Analysis-of-Surface-Waves (SASW) method is a new in situ testing method for determining shear wave velocity profiles of soil sites. The method is nondestructive, is performed from the ground surface, and requires no boreholes. Measurements are made at strains below 0.001 percent where elastic properties of the materials are independent of strain amplitude. The key elements in SASW testing are the generation and measurement of surface waves, Rayleigh waves. Two receivers are located on the ground surface and a transient impact containing a large range of frequencies is transmitted to the soil by means of a simple hammer. The surface waves are captured and recorded by the receivers using a spectral waveform analyzer. The analyzer is used to transform the waveforms into the frequency domain and then to perform spectral analyses on them. The points of interest from this operation are the phase information of the cross power spectrum and the coherence function. By evaluating the coherence function during testing, the range of frequencies which is not contaminated with random background noise can be quickly identified, so that the quality of the signals being saved for further data reduction is insured. Phase information from the cross power spectrum is indicative of the relative phase shift of each frequency propagating between the two receivers. By knowing the distance between receivers and the phase shift for each frequency, phase velocity and wavelength associated with that frequency are calculated. With this information a dispersion curve can be constructed. A dispersion curve is a plot of phase velocity versus wavelength. By applying an inversion process, an analytical technique for reconstructing the shear wave velocity profile from the dispersion curve, layering and the shear wave velocity and shear modulus of each layer can be readily obtained. One of the most important steps in SASW testing is the inversion process which has been the missing link in engineering applications.

Seven case studies are presented to illustrate the utility of the SASW method. Six of the sites experienced liquefaction in either the 1979 Imperial Valley or 1981 Westmorland Earthquakes. At one site shear wave velocities are compared with those of the crosshole seismic method, with the velocities comparing closely. At all sites where liquefaction occurred, in situ values of V are less than 450 fps. Such low shear wave velocities combined with saturated liquefiable soils represent a strong indicator of the likelihood of liquefaction.

ACKNOWLEDGEMENT

This work was supported by the Texas State Department of Highways and Public Transportation and the United States Geological Survey. The writers would like to thank Mr. R. B. Rogers and Mr. J. F. Daleiden from TSHPT and Dr. L. T. Youd from USGS.

REFERENCES

1. Ballard, R.F., Jr. (1964), "Determination of Soil Shear Moduli at Depths by In Situ Vibratory Techniques," Miscellaneous Paper No. 4-691, U.S. Army Engineer Waterways Experiment Station, Vicksburg, MS.
2. Ballard, R.F., Jr., and Casagrande, D.R. (1967), "Dynamic Foundation Investigations, TAA-2A Radar Site, Cape Kennedy, Forida," Miscellaneous Paper 4-878, U.S. Army Engineer Waterways Experiment Station, Vicksburg, MS.
3. Dobry, R., Pawell, D.J., Yokel, F.Y., and Ladd, R.S. (1980), "Liquefaction Potential of Saturated Sand - The Stiffness Method," Proceedings, Seventh World Conference on Earthquake Engineering, Istanbul, Turkey.
4. Dobry, R., Stokoe, K.H., II, Ladd, R.S., and Youd, T.L. (1981), "Liquefaction Susceptibility from S-Wave Velocity," Proceedings, In Situ Testing to Evaluate Liquefaction Susceptibility, Preprint 81-544, ASCE Annual Convention, St. Luis, MI.
5. Gazetas, G. (1982), "Vibrational Characteristics of Soil Deposits with Variable Wave Velocity," International Journal for Numerical and Analytical Methods in Geomechanics, Vol. 6, pp 1-20.
6. Geological Survey (1982), "The Imperial Valley, California, Earthquake of October 15, 1979," Professional Paper 1254, U. S. Department of Interior, Washington, DC, 451 pages.
7. Grant, F.S., and West, G.F. (1965), Interpretation Theory in Applied Geophysics, McGraw-Hill Compony, NY, 584 p.
8. Haag, F.D., Stokoe, K.H., II, and Nazarian S. (1985), "Seismic Investigation of Five Sites in the Imperial Valley, California after the 1981 Westmorland Earthquake," Geotechnical Engineering Center, University of Texas, Austin, Texas.
9. Haskell, N.A. (1953), "The Dispersion of Surface Waves on Multilayered Media," Bulletin of Seismological Society of America, Vol. 43, No. 1, pp. 17-34.
10. Heisey, J.S., Stokoe, K.H., II, and Meyer, A.H. (1982), "Moduli of Pavement Systems from Spectral Analysis of Surface Waves," Transportation Research Record, No. 853, Washington, D.C.
11. Hewlett-Packard (1981), "The Fundamentals of Signal Analysis," Application Note 243, Hewllet-Packard Co., 57 p.
12. Nazarian, S., and K.H. Stokoe, II (1983), "Use of Spectral Analysis of Surface Waves for Determination of Moduli and Thicknesses of Pavement Systems," Transportation Research Record, No. 954, Washington, D.C.
13. Thomson, W.I. (1950), "Transmission of Elastic Waves through a Stratified Solid Medium," Journal of Applied Physics, Vol. 21, pp. 89-93.
14. Youd, L.T., and Weczorek, G.F. (1981), "Liquefaction and Secondary Ground Failure," U.S. Geological Society Professional Paper.

COMPUTER-AIDED STUDIES OF COMPLEX SOIL MODULI

by Robert D. Stoll[1] M. ASCE

ABSTRACT

Low-cost microcomputers and new electronic components make it possible to measure accurately the phase and amplitude of mechanical waves in the frequency domain at a small fraction of the cost that prevailed only a few years ago. Both laboratory and field experiments benefit from this capability and in this paper we present examples of both. In a series of new laboratory experiments we have measured the phase and amplitude of both the driving torque and the resulting torsional motion of a cylindrical specimen. Motion is sensed with a capacitive, noncontact probe and torque is measured by use of a torque beam instrumented with a silicon strain-gage bridge. The output from the bridge and the probe are channeled through a 12 bit A/D converter and then a discrete Fourier transform is used to determine the real and imaginary parts of each signal. In each experiment, a robust sample is taken and a sliding window is then used to increase the accuracy of phase measurements. The loss-angle obtained by comparing the phase of the driver and the response compares well with conventional measurements of logarithmic decrement at various resonances. The microcomputer and Fourier transform techniques are also used in the interpretation of some of our field work involving Rayleigh and Stoneley waves. In this case we digitize the data from two or more geophones and calculate the phase difference between phones at different frequencies to determine the phase velocity while the amplitude spectra are used to study attenuation.

INTRODUCTION

Accurate measurement of the phase difference between two related analog signals is a technique which allows the determination of several important physical properties of interest in soil dynamics and geophysics. For example in performing slow torsional tests, the ratio of the imaginary to the real part of the complex modulus $G^*(\omega) = G_r(\omega) + iG_i(\omega)$, is equal to the tangent of the loss angle, δ_L. Since the loss angle is the difference in phase between the driving torque and the resulting rotational motion, the complex modulus may be determined by careful measurement of these quantities followed by an analysis of the phase difference between them.

At higher frequencies, as various resonant modes are excited, the traditional methods of determining the damping involve either measuring the decay of free vibrations after a sample is driven to resonance, yielding the logarithmic decrement, δ, or measuring the width of the

---

[1]Professor of Civil Engineering, Columbia University, New York, New York 10027.

resonance curve at some arbitrary level. However, if careful measurements of phase difference are made, it is possible to determine the quality factor, Q, more accurately using a small portion of the data near the center frequency and simple narrow band filter theory (7). For small amplitudes of motion the complex modulus, the quality factor, the loss angle, and the logarithmic decrement are interrelated approximately such that (8)

$$1/Q = \delta/\pi = \tan \delta_L = G_i/G_r \qquad (1)$$

and from filter theory Q and $\eta(\omega)$, the phase shift near resonance, are also related by the expression

$$\eta(\omega) = \arctan\left(\frac{\omega/\omega_o - \omega_o/\omega}{Q^{-1}}\right) + C \qquad (2)$$

where ω is drive frequency, ω_o is the resonant frequency, and C is a constant that accounts for instrument phase-shift error. Thus by measuring the phase shift at various frequencies near resonance we may fit Eq. 2 to the data in order to determine Q.

There are also a number of situations in the field where phase measurements are very helpful. In using surface or interface waves, such as Rayleigh or Stoneley waves, to determine shear wave velocity, one traditional approach has been to move the motion sensor along the surface of the ground to determine the wavelength of motion generated by a vibrator. In underwater applications, it is generally not feasible to move the transducer so that measurement of the phase difference for a signal received at two fixed locations is a much more practical approach. Moreover, in the case where transient waves are used to measure in situ shear wave velocity, it is much more accurate to determine the phase difference of various frequency components rather than to work with "first breaks" in the time domain.

Information about the phase of a signal can be obtained in a number of different ways. Special analog equipment is available that will perform coherent demodulation of a received signal into its "in phase" and "quadrature" components with respect to some reference or source signal. Analog to digital conversion is then performed and the ratio of the two components is used to calculate the phase. Alternatively the source and received signals or a set of received signals may be sampled in the time domain, and then a discrete Fourier transform (DFT) may be performed to obtain the complex Fourier components at any desired frequency. The complex transform yields both amplitude and phase spectra and so may be used to study phase differences in a variety of different applications. In using this approach, the need for expensive analog equipment is largely avoided and there is a great deal of flexibility in the way in which the signal may be processed to reduce noise and enhance the features that are of interest. Moreover, with the advent of modern microcomputers and relatively inexpensive A/D and D/A converters, the entire sequence of sampling, filtering, Fourier-transforming, interpreting and displaying the results can be done on a single microcomputer at a very modest cost.

We have followed the latter approach and have developed a number of signal processing techniques that utilize a microcomputer for control and processing of both laboratory and field data. In this paper examples of both are presented.

BASIC EQUIPMENT

Fig. 1 shows a block diagram of the equipment that is used in our experimental work. In most cases the analog input falls into one of the following categories:

1. AM or FM tape recorder output with field data from geophones or accelerometers used in seismic studies or studies of structural vibrations.

2. Transducer output from laboratory studies involving either pulse or harmonic loading. The following transducers are most commonly utilized:

    a) Capacitive or inductive probes that sense displacement without contact.

    b) Silicon or conventional strain-gage bridges used to measure torque, pressure, or force on an experimental component.

    c) Piezoelectric accelerometers or electromagnetic velocity sensors used to measure motion of various experimental components.

The analog filters, which are placed ahead of the A/D converter, serve the dual purpose of enhancing the signal to noise ratio by removing unwanted high and low-frequency components as well as removing all high frequency components which would cause aliasing in later signal processing. To prevent aliasing, the highest frequency included in the signal (the Nyquist frequency) must always be less than one half the frequency of sampling.

For most of our work we have used a 12-bit A/D converter with eight channels of multiplexed differential input. The present unit plugs directly into one of the expansion slots in the microcomputer and allows direct memory access (DMA). The DMA feature allows large blocks of data to be acquired efficiently and placed in memory without burdening the central processor with the tasks of timing and control of input and output signals. Moreover, the sampling rate is determined by a separate crystal-controlled clock and hardware dividing circuits so that no machine language routines are necessary to control the multiplexing and sampling as was required in our earlier systems. Sampling rates of up to 37 kHz can be handled by the present unit. In an earlier version of our equipment, the computer was required to do such "housekeeping" chores as controlling I/O and multiplexing so that the assembly language routines had to be carefully written to be sure that the precise time of each sampling event in each channel was known since we are interested in measuring the small phase differences that will occur from channel to channel.

Two independent channels for D/A conversion are also available

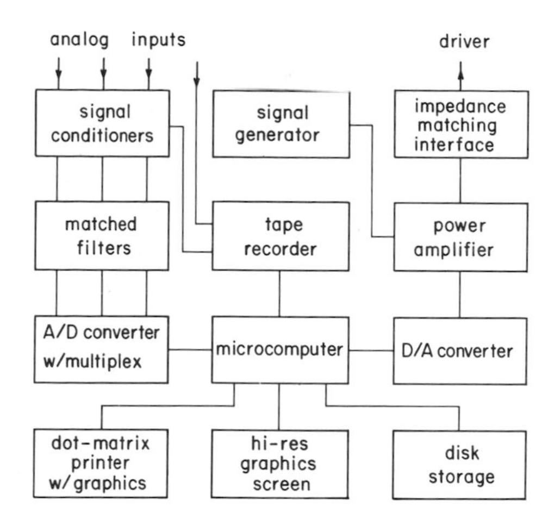

Fig. 1. Components used in experimental work.

on our present system so that the computer can be used to drive an experiment with an arbitrary, preprogrammed signal at the same time that the response is being recorded. In the earlier version of our equipment which contained no D/A facilities, an ordinary signal generator was used to drive the experimental apparatus.

The microcomputers used in our experiments have evolved over the past several years from an 8-bit unit with a maximum of 64K of memory (Radio Shack and LNW) to an 8/16 bit unit with over 512K of memory (IBM). The units used earlier in our work are quite adequate for the work described herein; however, the expanded memory of the IBM as well as many other newer micros is very useful in handling many experimental records simultaneously for graphics displays, correlation, etc. We use the high-resolution screen graphics to display data points, pick data windows, and visually compare different signals prior to processing. The dot-matrix printer allows the data to be plotted for a permanent record and the results at various stages of the signal processing to be recorded graphically, both for archiving and for presentation or publication in a report.

TIME-TO-FREQUENCY TRANSFORMATION

Since most of our data is transformed to the frequency domain for part of the analysis, some comment on the use of the Fourier transformation is helpful at this point. When an analog signal is sampled in the time domain over a "window" of width T (sec), the following transformation may be used to determine the coefficients, A_r, in the frequency domain

$$A_r = \sum_{k=0}^{N-1} X_k \exp(-i2\pi rk/N) \tag{3}$$

where N is the number of data points and X_k is the amplitude of the signal being sampled. If r is chosen to be the set (r=0,1,....,N-1) then Eq. 3 gives what is usually termed the discrete Fourier transform (DFT) of the sampled data. Moreover, when one of several particularly efficient computational algorithms is used to calculate the series given by Eq. 3, the result is called the fast Fourier transform (FFT). The FFT algorithm simply takes into account the many repetitions in numerical value that occur in the sine and cosine functions used to evaluate the complex exponential in the transform. When calculating the DFT, the frequency corresponding to each value of r is given by the expression

$$f_r = r/T = r/N\Delta t \tag{4}$$

where Δt is the sampling interval. Since N is usually chosen to be a multiple of 2, Δf, the frequency interval, is a predetermined fractional value which is often awkward to work with and seldom coincident with the frequencies that are normally chosen in experimental work. Thus it is sometimes necessary to give up the speed of the FFT and choose a different set of values for r. The resulting "brute force" Fourier transform (FT) may be structured to give Fourier coefficients at any desired frequency for the same data set used in the DFT or FFT. The tradeoff, however, is that the calculations require much more computer time. An example of the 256 point FFT and a portion of an FT centered around the principal peak in the spectrum is given in Fig. 2. In Fig. 2 the sampling interval was 2.5 msec so that the frequency interval for the FFT is 1.56 Hz. By arbitrarily choosing the frequency interval to be .156 Hz and performing an FT, we obtain the portion of the spectrum shown in the insert labeled "high resolution near peak amplitude". The interesting point here is that the largest coefficient given by the FFT is "off-channel" compared to the peak amplitude revealed by the FT. In some experiments we wish to compare phase angle at or very near the "on-channel" or peak amplitude and so the full FT computation becomes necessary. An excellent discussion of the practical use of transforms is given in Ref. 5.

In any transform computed from data sampled over a finite window, the spectrum will contain features that depend on the type of sampling window that is used. For example, if vibrations produced by a source of fixed frequency are sampled, the larger the number of cycles included in the data window the more nearly the spectrum will simulate the single spike that would result from a continuous Fourier transform obtained analytically. Moreover, if data near both ends of the window are filtered to avoid a sudden step to zero, much of the so-called leakage and resulting distortion can be eliminated from the discrete transform.

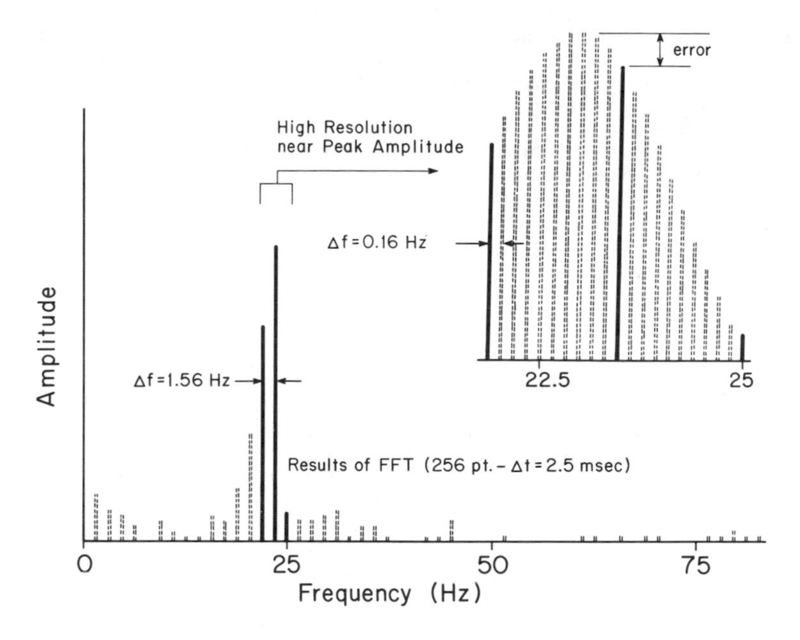

Fig. 2. Fourier spectrum for FFT and for high resolution FT.

LABORATORY EXPERIMENTS

As a first example of the use of a microcomputer in the laboratory, experiments designed to measure the damping in sediments at very low frequencies will be described. This work is part of a detailed study of long range, low-frequency wave propagation in marine sediments. In particular we were interested in extending earlier data obtained by utilizing the resonant column method, into the very low frequency range (one to two Hz) while still measuring at the lowest possible levels of shear strain. A number of authors have shown that the dynamic shear modulus tends towards a maximum and damping towards a minimum value as the shear strain amplitude becomes smaller than about 10^{-6} rad. In the very low frequency range, a conventional resonant column experiment, where velocity or acceleration transducers are used to measure the torsional motion, is not practical. Instead we employed a very sensitive noncontact capacitive transducer to measure the tangential displacement at the circumference of a specimen driven in torsion at very low frequencies. The torque actually transmitted to the specimen is measured at the point of application, and the difference in phase between the driving torque and the resulting motion is calculated after the data is transformed to the frequency domain as described earlier.

Part of the success of the experiments was due to the availability of a commercial, noncontact transducer that can resolve extremely small displacements. Such transducers are used commercially in manufacturing plants to sense the variation in thickness of thin sheets of different materials or to check the alignment of high-speed magnetic disks used for data storage in computers. The unit which we are using generates a radio frequency signal (3 MHz) at the tip of a probe positioned so that the measuring face is about .0025 mm from the moving target. The full scale output of the signal conditioner that is used with the probe is ± 10 volts at a displacement of ± .00025 mm, linear to within 0.4% of full scale output. Since we use the transducer in a range near its lower limit, extreme care in mounting the transducer and isolating the entire experimental apparatus from external noise caused by wind, air currents and structural vibrations is essential. Fig. 3 shows a schematic of one version of the experimental setup.

The torque applied to the top of the specimen is measured by sensing the strain in a torque beam attached at two points to the top of the specimen. A silicon strain gage bridge and DC amplifier provide the necessary electrical output for sampling by the A/D converter. During the test, alternate samples of the driving torque and the resulting torsional motion are made and stored directly in the computer's random access memory. In most cases a robust sample involving more than 100 cycles of motion is taken, after which a number of Fourier transforms are performed using a sliding window that is moved progressively through the data. A cosine taper is usually applied to 10% of the data at each end of the data window. Fig. 4 shows a plot of the data points from one such window together with the amplitude spectra from an FFT. In addition to the FFT, a full FT was performed at frequencies near the maximum spectral ordinate, as illustrated previously. The phase of the two signals was then obtained by taking the ratio of the imaginary and real parts of the transform at the frequency corresponding to the maximum amplitude and correcting for the precise offset caused by the multiplexing.

Each phase difference measured in the manner described above must be corrected for any instrument error that may change the phase of the signal both before and after it is sampled. For example, the low pass filter and the signal conditioning circuitry most certainly introduce some differential phase shifts no matter how carefully they are designed. This is particularly important in view of the fact that we are attempting to measure real phase differences of the order of one degree or less. For this reason the entire system, operating at essentially the same amplitude and frequency as for the real experiments, was calibrated by substituting a solid aluminum beam for the specimen. The insert in Fig. 3 shows a schematic of the calibration setup. Since the aluminum calibration beam has a damping ratio at least one order of magnitude less than the real specimen, one can use the response of the beam to determine the phase lag of the system and thus largely eliminate the instrument error.

Results of the low frequency experiments described above were combined with additional data obtained from more conventional resonant column experiments to obtain a continuous spectrum of loss angle over the frequency range from one or two Hz to over 1kHz for several different gradations of sand and for an inorganic silt. The results of these

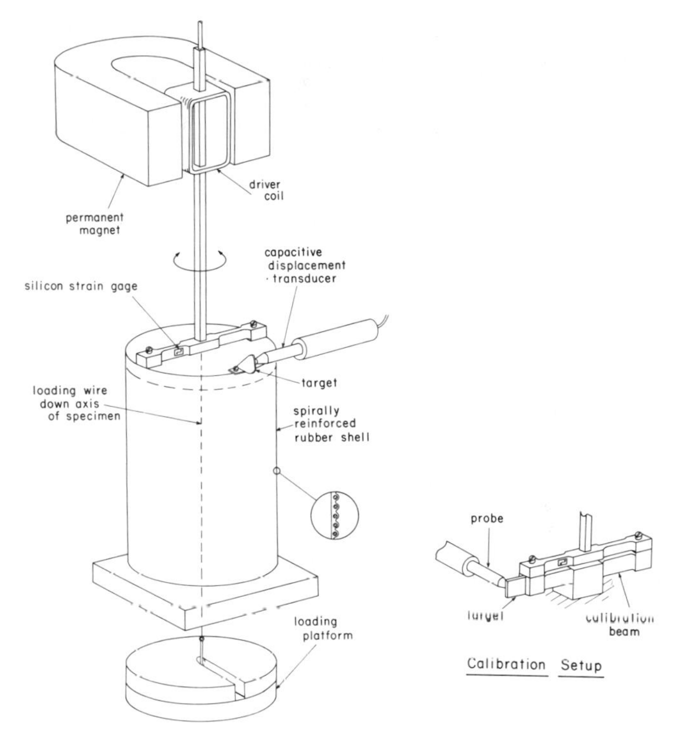

Fig. 3. Experimental apparatus used to measure loss angle.

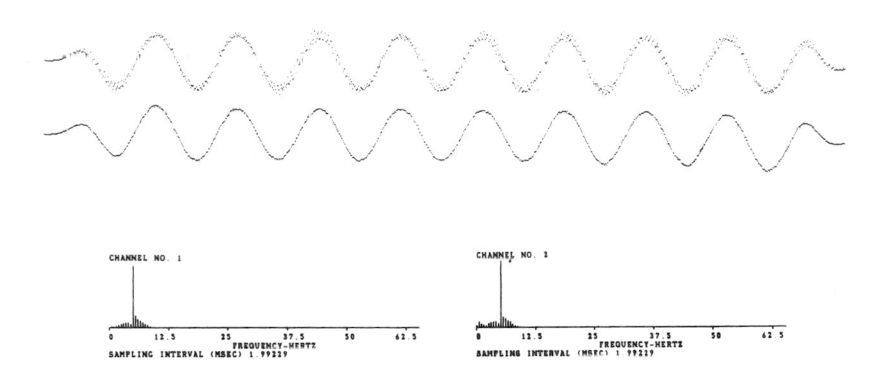

Fig. 4. Plot of data and Fourier spectra for one window - note taper at both ends of window.

experiments have been used to determine parameters in a mathematical model intended to describe acoustic wave propagation in water-saturated sediments. The model, which is based on the classical Biot theory (1-4), has been quite successful in predicting the results of laboratory and field experiments over a wide frequency range. One of the most important results of our work has been to identify several forms of viscous dissipation that tend to control the overall damping in water-saturated sediments at very low amplitudes of motion. The viscous damping overshadows the frictional losses at all but the lowest frequencies making the overall damping frequency dependent in saturated sediments at most frequencies of practical interest.

The apparatus shown in Fig. 3 is a modified version of equipment used previously (12) to measure resonant frequency and logarithmic decrement over a much narrower frequency range (30 Hz to 230 Hz). Specimens of water-saturated and dry sediment are confined in a columnar shape by thin rubber membrane which is reinforced with an embedded spiral of 0.25 mm diameter nylon filament wound with a spacing of appoximately 0.25 mm between turns. The shell produces stiff restraint to radial motion but very little resistance to either torsional or axial motion of the specimen. The top and bottom of the specimen are confined by rigid plates with rough faces, and thin stainless steel wire, which is attached to the top plate and runs down the axis of the specimen through an "O" ring gland in the bottom plate, is used to apply a static axial load. The driver mechanism shown in Fig. 3 was used for low-frequency tests; however, for higher frequencies where the specimen was driven through several different modes of resonance, lightweight coils were mounted directly on the top plate of the specimen. Further details of the apparatus are given in Ref. 12.

Results of a series of tests on Ottawa sand are shown in Fig. 5. Ottawa sand is a standard testing sand with uniform, rounded grains. A fraction passing the no. 20 and retained on the no. 30 sieves, with a mean grain diameter of 1.0 mm was used for these tests. All of the data points are for measurements at a maximum shear strain amplitude of less than 10^{-6} rad. The permeability of this fraction of Ottawa sand is between 1 and $5x10^{-6}cm^2$ depending on the relative density of the specimen. For a granular material with this high a permeability, the Biot theory predicts that there will be a marked increase in damping in the frequency range from 10 Hz to 1 kHz due to the motion of the fluid relative to the skeletal frame. Several examples of the predictions of the Biot theory for sands of different permeability have been given by Stoll (9-11).

In Fig. 5, the open and solid circles are for measurements of logarithmic decrement reported by Stoll (12) in 1979. The triangles and squares are new data which extend the range of frequencies covered from 2 Hz to over 1 kHz. The square symbols correspond to measurements of phase difference as described above, while the triangles are for new measurements of logarithmic decrement. All of the open symbols are for tests on oven-dried samples where the only mechanism producing damping is intergranular friction. As a result the loss angle, δ_L, is a constant over the entire frequency range. On the other hand, when the specimen is saturated with water, there is a marked increase in the loss

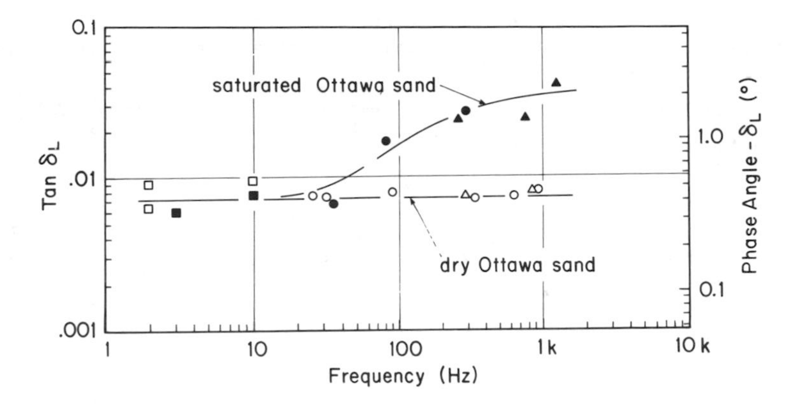

Fig. 5. Loss angle from experiments on 20-30 Ottawa sand.

angle starting at a frequency of about 10 Hz. Thus the test results suggest that a reasonable model for the dry skeletal frame may be based on the notion of a constant complex modulus since this will produce the required constant loss angle. Furthermore, when the sand is saturated, viscous losses caused by the fluid moving relative to the skeletal frame are adequately accounted for by the basic Biot theory which predicts a loss angle which is strongly dependent on frequency in a manner very close to what is seen in the figure.

The results of a second series of tests on water-saturated and dry silt are shown in Fig. 6. The silt used in these tests was an inorganic, micaceous material of glacial origin with a mean grain size of about .025 mm. As in the previous case, some of the data was obtained from prior tests where the resonant column technique was utilized. The results of a series of new tests involving both measurements of phase difference and logarithmic decrement have been added to the data set, and it can be seen that, for the saturated silt, the resulting pattern is consistent over the entire range of frequencies from one or two Hz to over one kHz suggesting that there is a gradually increasing effect of frequency up to the limit of our data. At a porosity of 0.5 the permeability of the silt that was used in these tests is about $2.5 \times 10^{-10} cm^2$; when the permeability is this low and constant complex moduli are used to describe the response of the skeletal frame, the Biot theory does not predict any appreciable frequency dependence of the loss angle until frequencies of the order of 10 kHz are reached (see Ref. 11). Thus the frequency dependence that is obvious from the data of Fig. 6 is apparently due to a different kind of damping mechanism than is contained in the basic Biot theory with only frictional losses reflected in the frame moduli.

On the basis of recent theoretical studies (13, 15), we attribute the frequency-dependent damping observed in this case to local viscous losses that occur near the points of intergranular contact. Oscillatory fluid motion near the points of contact is similar to "squeeze film" motion that is well known in lubrication theory, and the effect is to produce a resistance to the relative approach of particles that increases in proportion to frequency. This effect can be incorporated

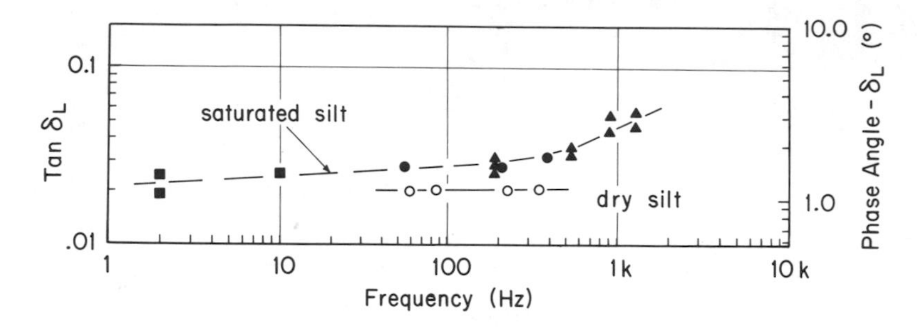

Fig. 6. Loss angle from experiments on inorganic silt.

into the Biot theory by assuming the response of the skeletal frame to be viscoelastic and defining a spectrum of relaxation times that will produce the right form of frequency dependence in the complex moduli that are used to describe the skeletal frame. Thus, in order to implement this approach, data covering a wide frequency range, such as shown in Fig. 6, is necessary in order to define the appropriate viscoelastic model. The current thrust of our experiment, work is to measure over an extended frequency range the real and imaginary parts of both G^* and Young's modulus, E^*, appropriate for a sediment in a water environment. With this information we can predict the propagation characteristics of p-waves in the low-frequency range on the basis of the acoustic model which we have developed over the past ten years (9-15). In order to measure E^*, we employ experiments in which extensional and flexural vibrations are studied using techniques similar to those discussed in this paper.

FIELD EXPERIMENTS

As a second example of the application of Fourier techniques, a field method for measuring the velocity of surface or interface waves will be described. In conventional studies of Rayleigh wave propagation, a surface transducer (geophone or accelerometer) is moved along the ground surface in an azimuthal path away from a source in order to determine the wave length of Rayleigh waves at different frequencies. For offshore, underwater work or in some areas of limited access, it is much more convenient to employ two transducers located at fixed locations and then sense the phase difference of the same signal at various frequencies. Here we face the opposite problem from the one encountered in the laboratory (where the phase differences are very small), since there may be differences in phase of more than a multiple of 2π between receiver points. Thus there is always a certain amount of ambiguity and some judgement or additional data is required in interpreting the data.

In order to be able to work in water of any reasonable depth from a single boat or platform, we have designed and constructed a gimballed, variable-frequency vibrator which can be lowered from shipboard together with a pair of gimballed geophones. The gimballed geophones are standard geophysical equipment designed for underwater or marsh environments. Fig. 7a is a picture of the vibrator and two different types of gimballed geophone. Fig. 7b shows the internal mechanisms of the vibrator. In the unit shown, the vibration is produced by a pair of counter-rotating weights driven by a variable speed DC motor. The entire mechanical unit, containing the rotating weights and motor, acts as a pendulum and rotates about the long axis of the vibrator so that it always produces a vertical force when the vibrator lies on a relatively flat surface. Another version of the vibrator with electromagnetic excitation is currently being designed. With this unit it will be possible to control both the force and the frequency.

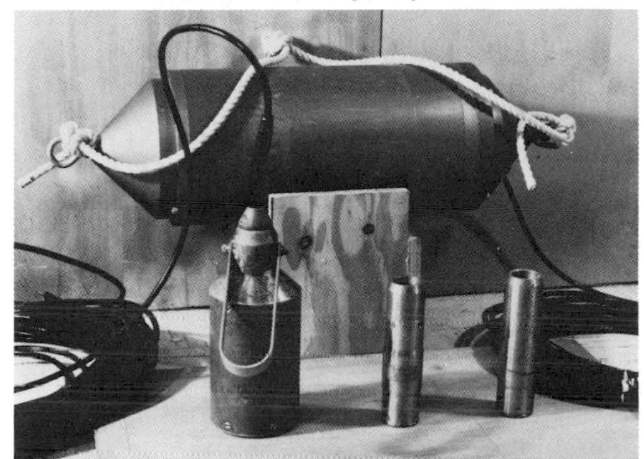

Fig. 7a

Fig. 7b

Fig. 7. Photographs of gimballed vibrator and geophones.

In addition to a gimballed vibrator, the source unit contains a pulse generator and a transmitting transducer to send high-frequency pulses when triggered. The pulse transducer, which can be seen at the right-hand end of the disassembled unit shown in Fig. 7b, is actually a piezoelectric hydrophone working in reverse. High-frequency pulses are radiated into the water and picked up by hydrophones that are lowered along with the geophones in order to determine the distance from the vibrator to the geophones. Signals received by the hydrophones and geophones are transmitted back to the source, by cable in shallow water or by FM radiobuoy in deeper water, to either be tape recorded or analyzed immediately. Fig. 8 shows a schematic of one of the field setups that is used to acquire data.

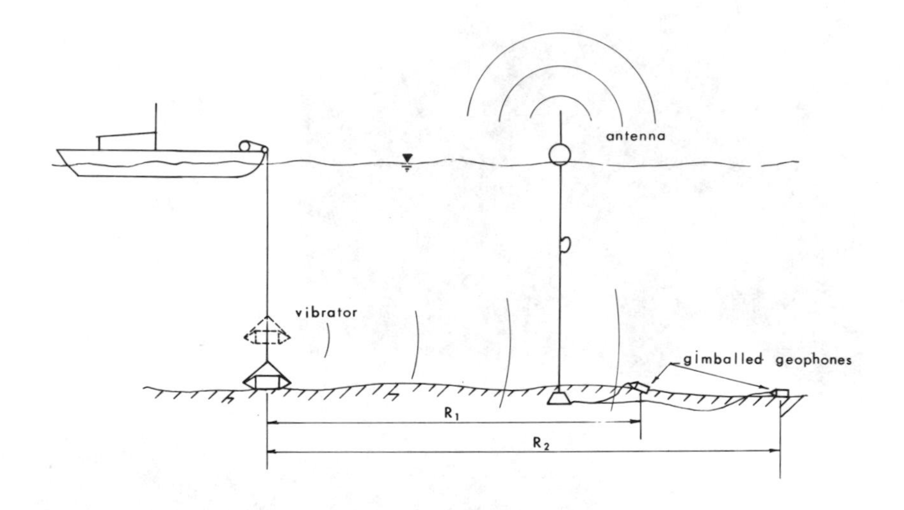

Fig. 8. Schematic of field setup.

A typical record of data that has been digitized and processed using the microcomputer is shown in Fig. 9. Since the distance between the two geophones is known, the phase velocity of the wave may be determined once the phase difference between the signals received at the two geophones is determined from the Fourier transforms. The inversion of the traveltime data to obtain velocity and attenuation as a function of depth in a nonhomogeneous deposit is not a trivial problem. However, recent theoretical analyses (6,16) have suggested a number of new approaches to this problem. Here again, the microcomputer may be used to good advantage in inverting the data and displaying the final results.

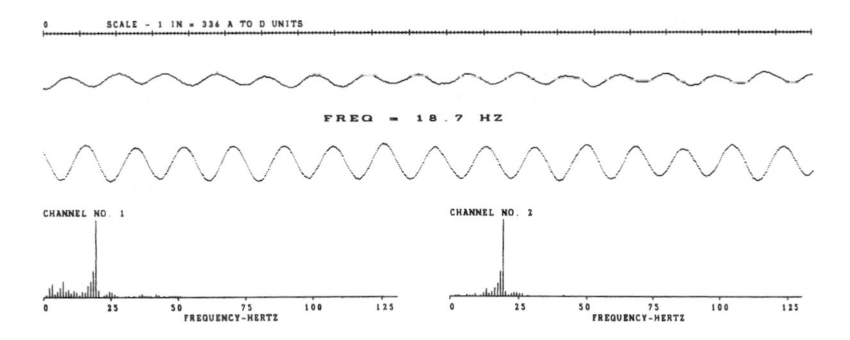

Fig. 9. Digitized data from two geophones and amplitude spectra from FFT.

SUMMARY AND CONCLUSIONS

Small integrated microcomputer systems that can control experiments, collect data, and perform analysis in an interactive manner have expanded the areas of research that are feasible in a low-budget university environment. The microcomputer can replace expensive, dedicated analog equipment as well as perform the tasks that were formerly possible only on full-size computers. By bypassing the large systems, where multiuser protocol makes it difficult to interface with real time experiments, the researcher can take advantage of many new and inexpensive interfaces that handle input and output to experimental components in a very efficient way. For example, by digitizing and collecting data directly in real time and storing it in computer memory, the need for storage on analog or digital tapes is eliminated. Moreover, the computer can be used to display and record the data set in any number of different ways, thus eliminating the need for a storage oscilloscope or other graphic displays. Finally, by performing various transformations such as the FFT and FT using software routines rather than expensive, hardwired analyzers, we eliminate the remainder of the traditional components that are used to perform many kinds of sophisticated experiments.

In this paper we have presented two examples of how such a system can be used to advantage in soil dynamics. In one case we have shown how accurate phase analysis can be utilized in laboratory studies of damping while, in a second example, it was shown how studies of wave propagation in the field can be analyzed by converting data to the frequency domain. In addition to these applications, the system described in this paper has also been used in a number of other experiments including studies of velocity and attenuation in refracted acoustic waves propagating through the ocean bottom (16). In each new application, the software base is expanded and the system becomes more and more versatile.

ACKNOWLEDGEMENTS

The work described in this paper was supported by the Office of Naval Research under contracts N00014-80-C-0098 and N00014-84-C-0132.

APPENDIX - REFERENCES

1. Biot, M.A., "Theory of Elastic Waves in Fluid-Saturated Porous Solid. I. Low Frequency Range," J. Acoust. Soc. Am., Vol. 28, 1956a, pp. 168-178.

2. Biot, M.A. "Theory of Elastic Waves in Fluid-Saturated Porous Solid. II. Higher Frequency Range," J. Acoust. Soc. Am., Vol. 28, 1965b, pp. 179-191.

3. Biot, M.A. "Mechanics of Deformation and Acoustic Propagation in Porous Media," J. Appl. Phys., Vol. 33, 1962a, pp. 1482-1498.

4. Biot, M.A., "Generalized Theory of Acoustic Propagation in Porous Media," J. Acoust. Soc. Am., Vol. 34, 1962b, pp. 1254-1264.

5. Higgins, R.J., "Fast Fourier Transform: An Introduction with some Minicomputer Experiments," Am. J. Phys., Vol. 44, 1976, pp. 766-773.

6. Nazarian, S. and Stokoe, K.H., "Nondestructive Testing of Pavements using Surface Waves," Transportation Research Record, 1984.

7. O'Brien, M.H. et al., "Digitally Controlled Measurement of some Elastic Moduli and Internal Friction by Phase Analysis," Rev. Sci. Instrum., Vol. 54, 1983, pp. 1565-1568.

8. Richart, F.E., Hall, J.R., and Woods, R.D., Vibrations of Soils and Foundations, Prentice Hall, Englewood Cliffs, N.J., 1970, pp. 158-167.

9. Stoll, R.D., "Acoustic Waves in Saturated Sediment," in Physics of Sound in Marine Sediments, edited by L. Hampton, Plenum, N.Y., 1974, pp. 19-39.

10. Stoll, R.D., "Acoustic Waves in Ocean Sediments," Geophysics, Vol. 42, 1977, pp. 715-725.

11. Stoll, R.D., "Damping in Saturated Soil," Proc. Specialty Conf. on Earthquake Engrg. and Soil Dynamics, ASCE, New York, 1978, pp. 969-975.

12. Stoll, R.D., "Experimental Studies of Attenuation in Sediments," J. Acoust. Soc. Am., Vol. 66, 1979, pp. 1152-1160.

13. Stoll, R.D., "Theoretical Aspects of Sound Transmission in Sediments," J. Acoust. Soc. Am., Vol. 68, 1980, pp. 1341-1350.

14. Stoll, R.D., "Computer-Aided Measurements of Damping in Marine Sediments," Proc. 2nd Internat. Conf. Computational Methods and Experimental Measurements, ISCME, Southampton, England, 1984, pp. (3)29-(3)39.

15. Stoll, R.D., "Marine Sediment Acoustics," J. Acoust. Soc. Am., 1985, (in press).

16. Stoll, R.D., and Houtz, R.E., "Attenuation Measurements with Sonobuoys," J. Acoust. Soc. Am., Vol. 73, 1983, pp. 163-172.

17. Vardoulakis, I., "Surface Waves in a Half-Space of Submerged Sand," Earthquake Engrg. and Struct. Dynamics, Vol. 9, 1981, pp. 329-342.

SEISMIC CPT
TO MEASURE IN-SITU SHEAR WAVE VELOCITY

P.K. Robertson[1], M.ASCE., R.G. Campanella[2], M.ASCE.,
D. Gillespie[3] and A. Rice[4]

Abstract

A new test, called the seismic cone penetration test (SCPT) is described. A small rugged velocity seismometer has been incorporated into an electronic cone penetrometer. The combination of the seismic downhole method and the CPT logging provide an extremely rapid, reliable and economic means of determining stratigraphic, strength and modulus information in one sounding. Results using the seismic cone penetration test are presented and compared to conventional in-situ techniques.

Introduction

In the last two decades, there has been an increasing interest in soil dynamics, mainly due to the many engineering problems in which the dynamic behaviour of soil is of significance. The increasing interest has resulted in the rapid development of new analytical and dynamic testing methods.

In the field the cross-hole and down-hole methods have become the standard techniques for dynamic testing to determine the in-situ shear wave velocity. A polarized shear wave is generated in one borehole (or at the surface) and the time is measured for the shear wave to travel a known distance to the geophone in the borehole. Elastic theory relates the shear modulus, G, soil density, ρ, and shear wave velocity, V_s as follows,

$$G = \rho \; V_s^2 \tag{1}$$

Hence, the shear modulus can be determined using in-situ seismic methods for the determination of the shear wave velocity. The shear modulus is largest at low strains and decreases with increasing shear strain (9). The shear strain amplitude in in-situ seismic tests is usually low and of the order of 10^{-4}%. Thus, the very low strain level dynamic shear modulus, G_{max} is usually obtained. The cost of such a

[1] NSERC University Research Fellow, Dept. of Civil Engineering, Univ. of B.C., Vancouver.
[2] Professor, Dept. of Civil Engineering, Univ. of B.C., Vancouver.
[3] Graduate Research Assistant, Dept. of Civil Engineering, Univ. of B.C., Vancouver.
[4] Engineer, Golder Associates, Vancouver.

test is usually high because of the requirement to have one or more boreholes. This has generally made the technique difficult for off-shore use.

A new test, called the seismic cone penetration test (SCPT), which can dramatically reduce the cost associated with the in-situ determination of shear wave velocity, will be described in this paper.

Seismic CPT

The Cone Penetration Test (CPT) is already used extensively off-shore and on-shore for geotechnical investigations. A cone of 10 cm^2 (1.55 in^2) base area with an apex angle of 60° is generally accepted as standard and has been specified in the European and American Standards (ASTM, 1979). A friction sleeve, located above the conical tip, has a standard area of 150 cm^2 (23.2 in^2). A pore pressure transducer has recently been added to measure the dynamic pore pressures during penetration. The cone penetrometer is pushed at the standard rate of 2.0 cm/sec (0.79 in/sec). Standard 1 m (3.28 ft) long rods are used to push the cone penetrometer into the soil. A cable, prethreaded through the center of the hollow push rods, connects the cone to the data acquisition system at ground surface. A small slope sensor is usually incorporated to monitor sounding verticality.

Full details of the design of an electronic cone are given by Campanella and Robertson (3). An example of a modern electronic cone penetrometer which also includes a temperature sensor is shown in Fig. 1. The piezometer-CPT is regarded as the premier test for the continuous logging of soil stratigraphy and shear strength. An example of the extensive data obtained from a piezometer-CPT is shown in Fig. 2. The cone data can be interpreted to give a good continuous prediction of soil type and shear strength (8). Predictions of soil stiffness (modulus) from the cone resistance can be rather poor, especially for overconsolidated soils, with a large potential error. The introduction of seismic measurements into the cone penetration test procedures enables the specific determination of the dynamic shear modulus (G_{max}). Thus, the combined measurement of soil strength and modulus provides a means to identify the complete stress-strain relationship of soil units which is required for many complex finite element analyses.

To obtain the measurement of dynamic shear modulus, a small rugged velocity seismometer has been incorporated into the cone penetrometer. The miniature seismometer is a Geospace GSC-14-L3 (1.7 cm dia) with a nominal natural frequency of 28 Hz. The seismometer is placed in the horizontal direction and orientated transverse to the signal source to detect the horizontal component of the shear wave arrivals (see Fig. 1). A schematic diagram showing the layout of the standard downhole technique is shown in Fig. 3.

A suitable seismic signal source should preferentially generate large amplitude shear waves with little or no compressional wave component. The shear waves travel through the soil skeleton and are thus related to the soil shear modulus. Results indicate that an

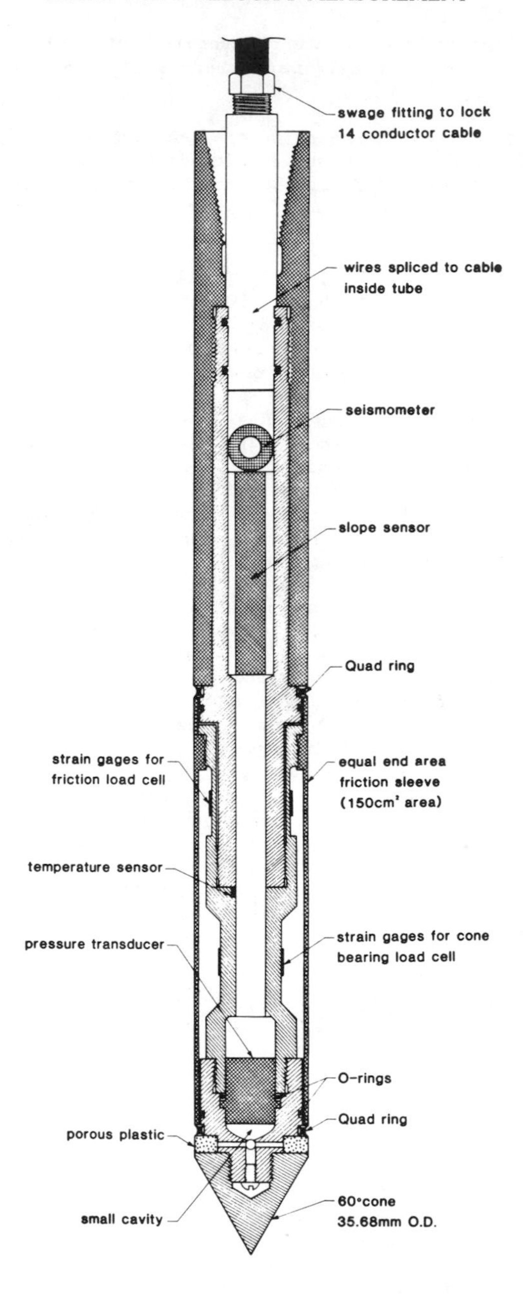

Fig. 1. Seismic Cone Penetrometer.

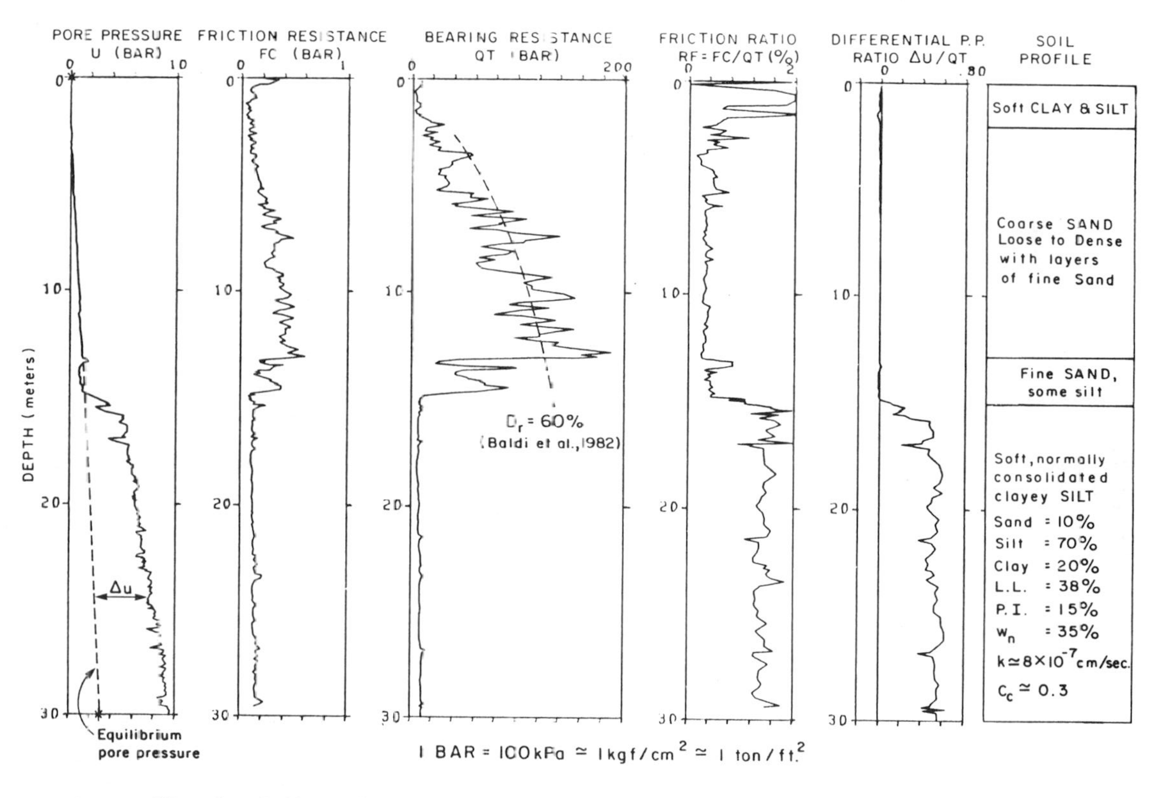

Fig. 2. Soil Profile for Research Site at McDonald's Farm Site (After Campanella et al. 1983).

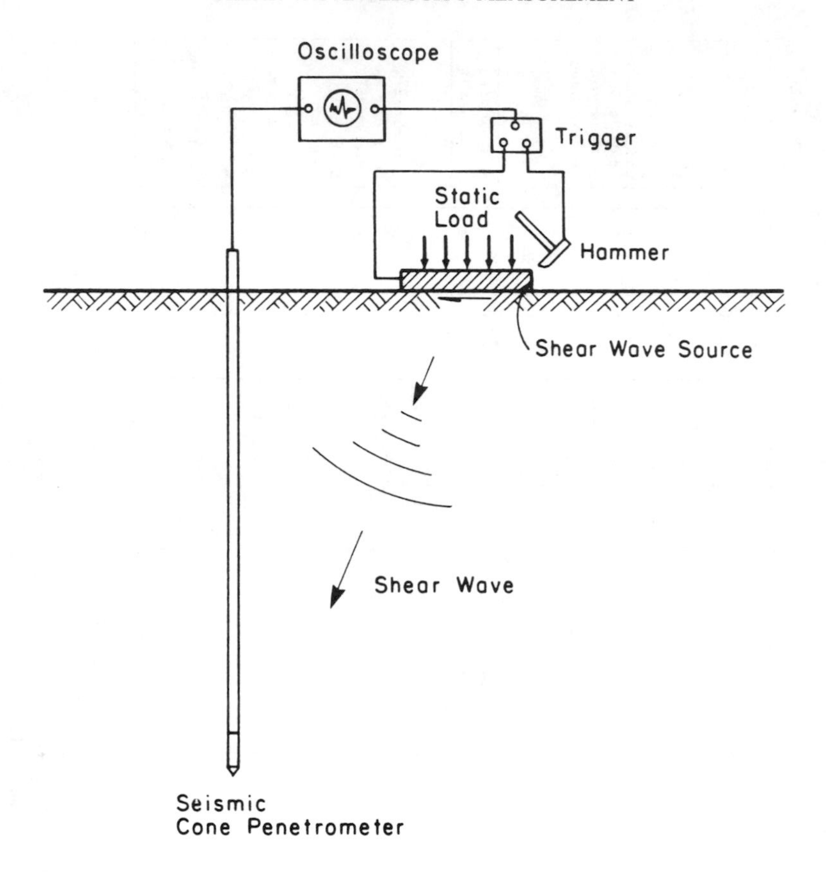

Fig. 3. Schematic Layout of Downhole Seismic Cone Penetration Test.

excellent downhole seismic shear wave source consists of a rigid beam or platform, steel jacketed and weighted to the ground. It may be struck with a sledge hammer as shown in Fig. 3. If the cone is being pushed by a drill-rig the beam can be weighted down by the rear pads of the drill-rig. If the cone is being pushed by a cone penetration vehicle, the beam can be weighted down by the pads of the vehicle or incorporated into the stabilizing pads for the truck. The beam type signal source is usually placed with ends equidistant within about 3 meters (10 ft) of the cone hole. The beam should be rigidly placed on the ground so that no energy losses should occur due to plastic shearing of the soil beneath the beam.

The design and construction of the seismometer carrier provides a snug seating for the seismometer package. The method of advancing the cone penetrometer provides continuous firm mechanical contact between the seismometer carrier and the surrounding soil. This allows excellent signal response. In addition, seismometer orientation can be controlled and accurate depth measurements obtained.

The seismic wave traces detected by the seismometer are recorded on a Nicolet 4094 digital oscilloscope with floppy disk capability. This unit has a 15 bit analog to digital signal resolution, very accurate timing capability and trigger delay capacity. The high resolution oscilloscope is capable of recording clean shear wave traces from forward and reverse single hammer impulses to depths of over 40 metres (131 ft), as shown in Fig. 4. Fig. 4 provides a quantitative comparison of the geophone response amplitude and relative shear wave travel times with depth. The geophone output voltage is directly related to the particle oscillation velocity as shown on the inset scales.

The strain level caused by the shear waves can be estimated at any depth during the CPT downhole seismic survey. The relationship between shear strain, γ, shear wave velocity, V_s, and peak oscillation velocity, u, is given by (10),

$$\gamma = \frac{u}{V_s} \tag{2}$$

Analysis of the existing field data shows that the strain amplitudes caused by the hammer-beam source are generally less than 10^{-4}% and decrease with depth.

It has been found that the time for the first cross-over point (shear wave changes sign) is easily identified from the polarized waves (forward and reverse) and provides the most repeatable reference arrival time. The arrival time from source to detector is converted vectorially to a vertical travel path. The difference between successive 1 m (3.28 ft) depth measurements of vertical travel path time is used to determine the shear wave travel time over the 1 m (3.28 ft) interval of depth. Because of the short distances and small travel times involved, the oscilloscope must have very high resolution, fast sample times and a very fast, repeatable trigger. The trigger used is similar to that suggested by Hoar and Stokoe (6). The trigger incorporates a MC1455 linear integrated circuit with a rise time of less than 1 μsec. Since the shear wave velocity is squared to calculate G_{max}, a high priority must be given to the accuracy of travel time measurements.

To assess the variability of the arrival time measurements and the accuracy and reliability of the trigger system a second geophone system was placed 1 m (3.28 ft.) vertically above the first geophone. This equipment modification allowed true interval surveys to be carried out and compared to the pseudo time interval method described above. The arrival time data was then analyzed assuming a normal statistical distribution at each depth interval. The results from a typical survey with 40 hammer blows using matched geophones is shown on Fig. 5. The data presented in Fig. 5 shows that there is very little error using a

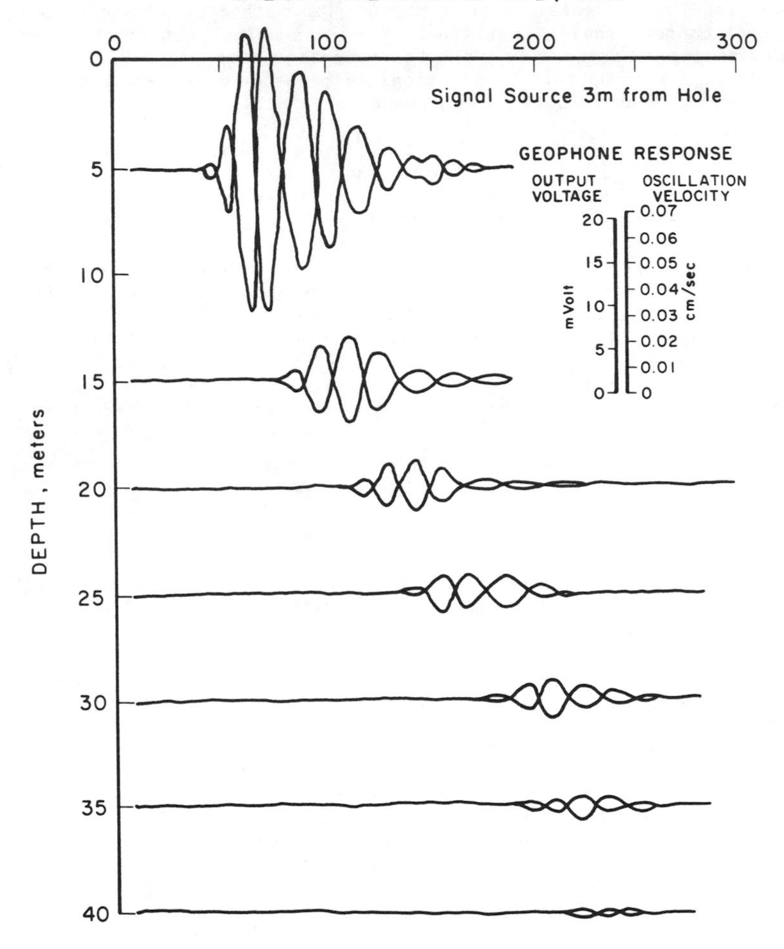

Fig. 4. Quantitative Comparison of Geophone Response Amplitude and Relative Shear Wave Travel Times from Seismic Cone Penetration Test.

single geophone and the pseudo time interval method as compared to the true time interval method with a pair of geophones.

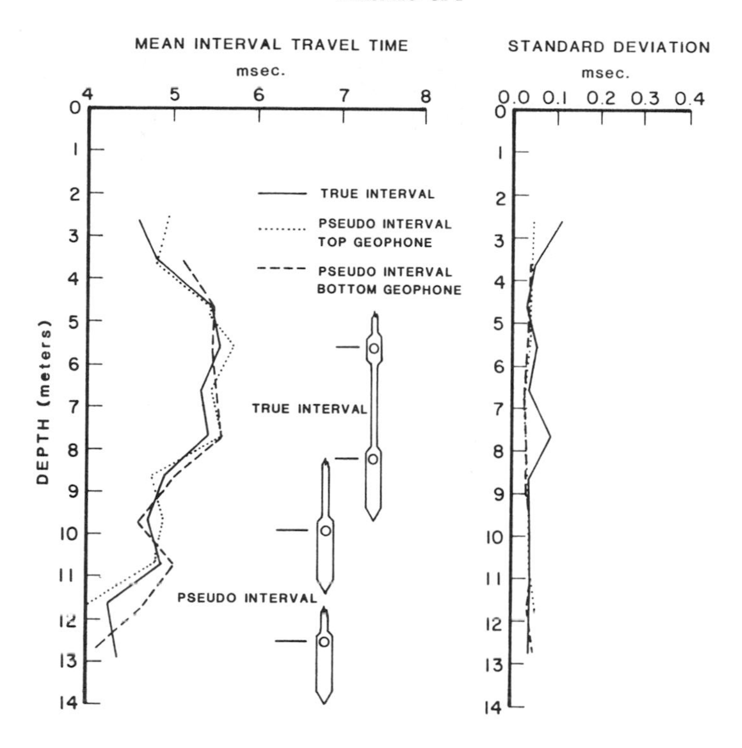

Fig. 5. Comparison of True and Pseudo Interval Travel Times.

Seismic CPT Results

Downhole seismic shear wave velocity measurements have been made at several sites and in some cases compared to results obtained by others using the conventional crosshole techniques. The seismic cone penetrometer was pushed into the ground at a constant rate of 2 cm/sec (0.79 in/sec). At approximately 1 m (3.28 ft) intervals, the penetration was stopped and shear waves generated at the surface by hitting a beam with a sledge hammer. Generally, only one blow with the hammer was required at each end of the plank to produce a single set of polarized shear waves.

McDonald's Farm Site, Vancouver

A research site for in-situ testing is located on an abandoned farm (McDonald's Farm) near the Vancouver International Airport. Full details of the site are given by Campanella et al. (4). A summary of the soil profile based on sampling and laboratory testing and cone penetration testing is shown in Fig. 2. The interval vertical shear wave velocities calculated from the difference of arrival times are shown in Fig. 6. Note that the results in Fig. 6 indicate that the interval shear wave velocity, and therefore maximum shear modulus, increases with depth. Also, the rate of increase with depth is higher in the sand than in the silt. Unfortunately, cross-hole seismic data does not, as yet, exist for this site, however, the maximum shear modulus calculated using the measured shear wave velocities shown in Fig. 6 compared extremely well with G_{max} values estimated from empirical relationships proposed by Seed and Idriss (9).

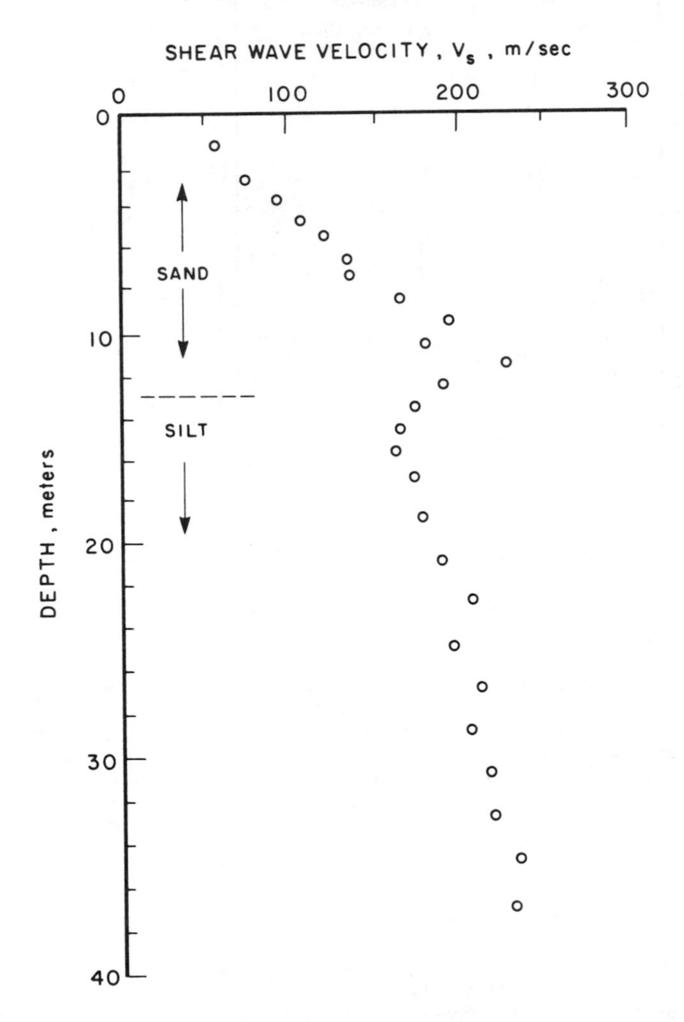

Fig. 6. Calculated Shear Wave Velocity Profile from Seismic CPT at McDonald's Farm Site, Vancouver.

Annacis Island, Vancouver

Extensive geotechnical investigations were carried out at the site of the new Annacis Bridge project near Vancouver. The area around the north main pier of the proposed cable stayed bridge consists of Fraser River sands to a depth of about 40 m (131 ft). The water table fluctuates with river level but is nominally about 4 meters (13.1 ft) below ground level.

A summary of the interval shear wave velocities and the cone bearing from the seismic CPT is shown on Fig. 7. The CPT seismic downhole profile was carried out approximately 5 meters (16.4 ft) from a three hole array used for a conventional crosshole seismic survey, which was carried out by others for the B.C. Ministry of Transportation and Highways. The crosshole data was obtained at 2.5 meter (8.2 ft) intervals and is also shown on Fig. 7. The CPT downhole data lies consistently above the crosshole data but generally the two sets of data compare within 20 percent. The seismic CPT data generally follows the trend indicated by the cone bearing profile with little in the way of dramatic velocity changes. The most notable changes occur at 4 m

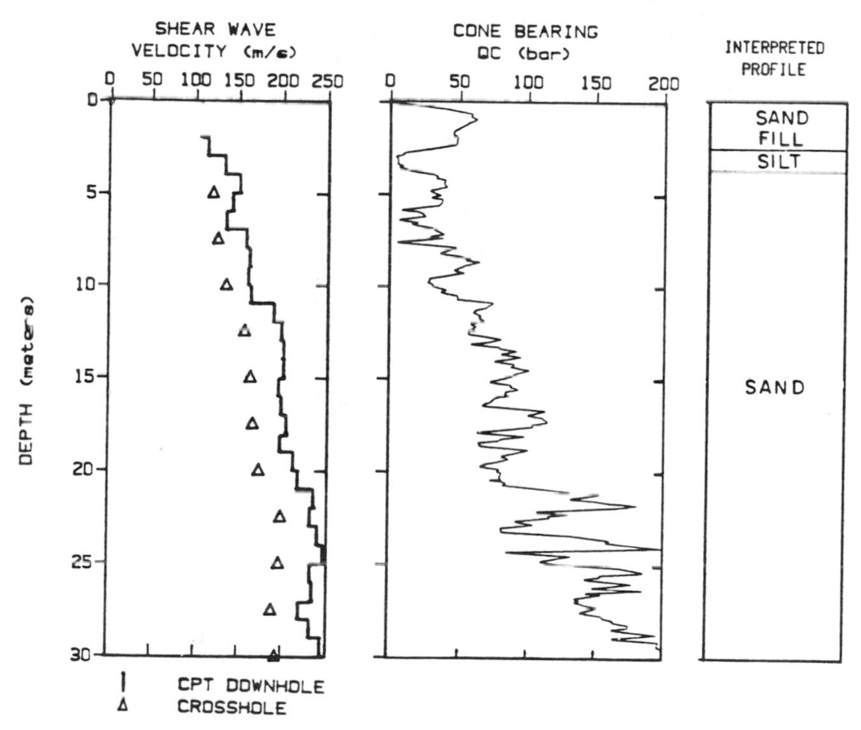

Fig. 7. Comparison of Seismic CPT Downhole and Crosshole Data at Annacis Site, Vancouver.

(13 ft) and 11 m (36 ft) where the shear wave velocities increase corresponding to a noticable increase in the cone bearing. The crosshole shear wave velocities, on the other hand, produce more subdued average values.

Imperial Valley, California

In spring 1984 seismic CPT tests were performed at several sites in the Imperial Valley, California with the cooperation of the U.S. Geological Survey and Purdue University. These sites were subjected to recent earthquakes; Imperial Valley, 1979 and Westmoland, 1981. Fig. 8 presents the seismic CPT data from the Wildlife site. Full details of the site are given by Bennett et al. (2). The wildlife site is located next to the Alamo River and exhibited extensive liquefaction during the 1981 earthquake. Also included in Fig. 8 is the shear wave velocities determined by crosshole tests (7). The two seismic profiles compare very favourably with velocities from the two independent methods generally within about 20 percent. Again, it is interesting to note that the shear wave velocities from the seismic CPT respond well to small variations in the soil profile.

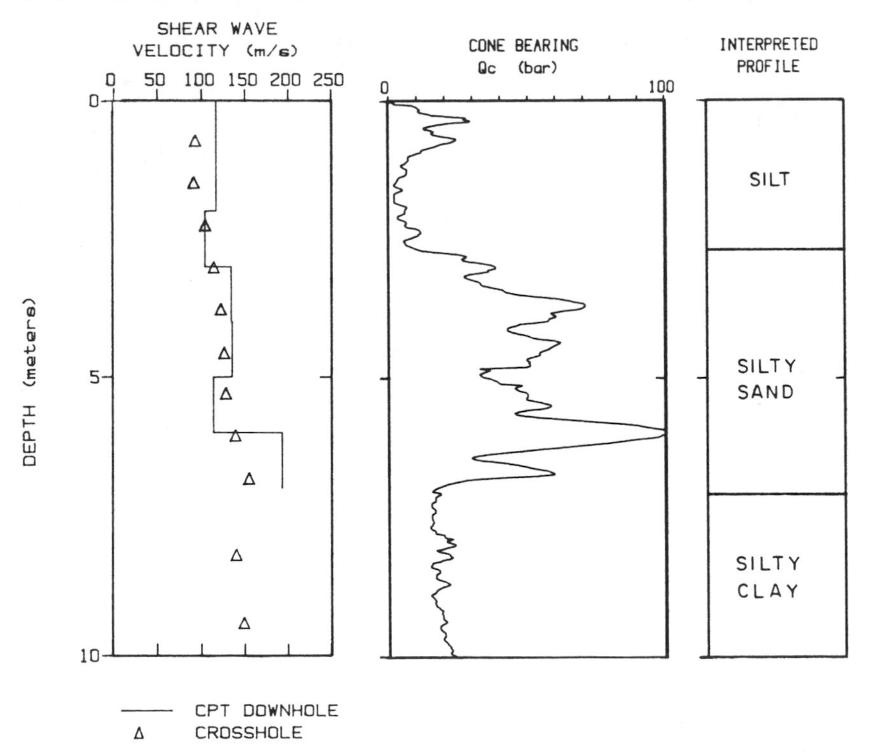

Fig. 8. Comparison of Seismic CPT Downhole and Crosshole Data at Wildlife Site, Imperial Valley.

Holmen and Museumsparken Sites, Drammen, Norway

In the fall of 1984 seismic CPT tests were performed at several sites in Norway with the cooperation of the Norwegian Geotechnical Institute (NGI) (5). These sites are well documented sites with extensive field and laboratory data.

The Holmen site consists of loose, medium to coarse sand to a depth of 25 m. Fig. 9 shows the seismic CPT data compared to the adjacent crosshole data. On average the two seismic shear wave velocity profiles are almost identical at this site showing little, if any, discrepancy between the CPT downhole and conventional cross-hole. The seismic CPT data responds well to variations in the soil profile observed from the cone bearing.

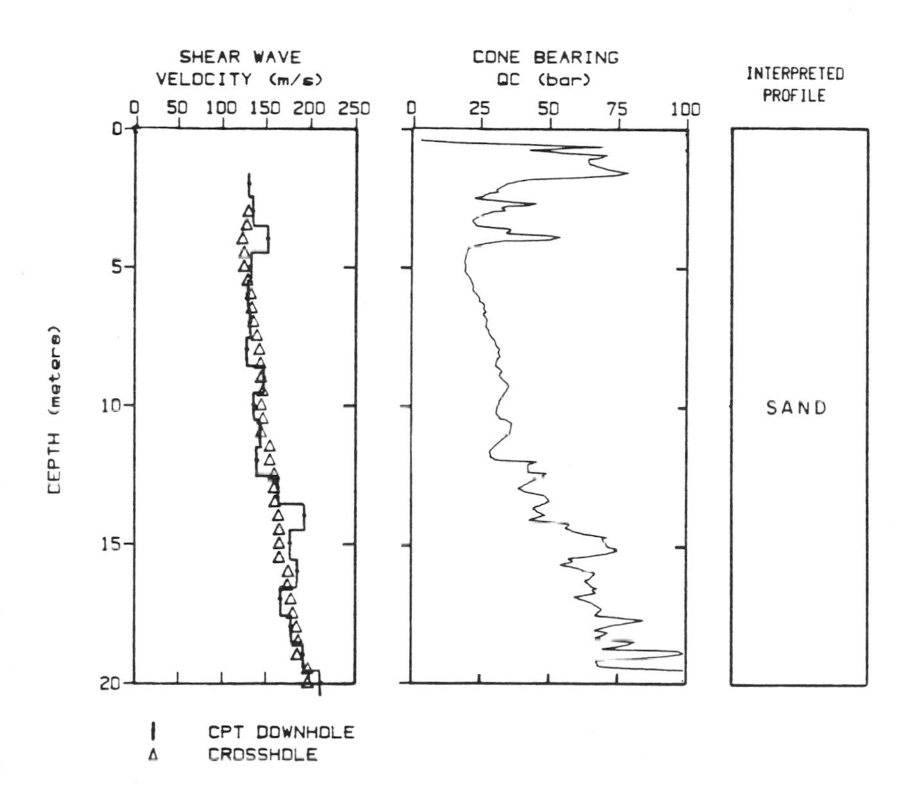

Fig. 9. Comparison of Seismic CPT Downhole and Crosshole Data at Holmen Site, Norway.

The Museumsparken site consists of the well documented Drammen clay to a depth of 15 m. Fig. 10 shows the seismic CPT data compared to adjacent crosshole data. Again the two seismic profiles compare very favourably.

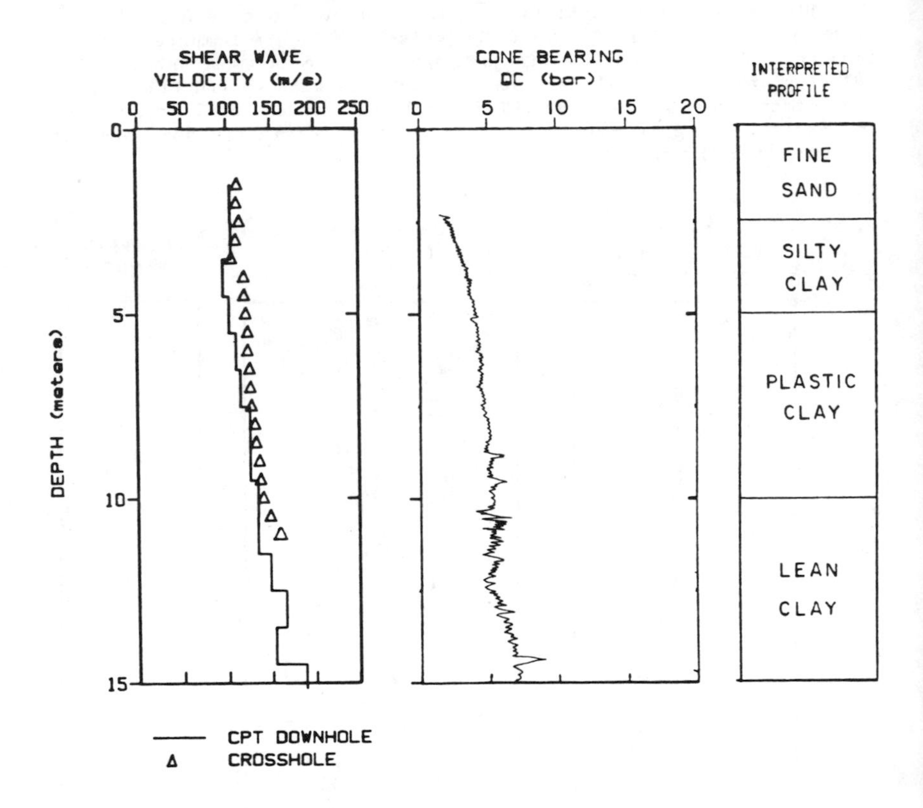

Fig. 10. Comparison of Seismic CPT Downhole and Crosshole Data at Museumsparken Site, Drammen, Norway.

Summary

A new test, called the seismic cone penetration test (SCPT) has been described. The cone bearing, friction sleeve stress and cone pore pressure data can be used to provide a fast and reliable determination of soil type and shear strength. Downhole seismic shear wave velocity measurements can be made during brief pauses in the cone penetration. The shear wave velocity data can be used to provide reliable determination of the maximum dynamic shear modulus. Accurate depth determination is made by measuring the rod length and seismometer orientation is easily maintained throughout the sounding. Hole verticality is monitored throughout the sounding with a small slope sensor installed in the cone. The combination of the seismic downhole method and the CPT logging provide an extremely rapid, reliable and economic means of determining stratigraphic, strength and modulus information in one sounding.

Comparison of the seismic CPT downhole shear wave velocity measurements with those obtained by conventional crosshole techniques show excellent agreement. The seismic CPT is, however, considerably less expensive and a more rapid procedure than the crosshole technique. The seismic CPT shows particular promise for use off-shore where CPT equipment is already used extensively and where shear wave velocity measurements would be of value in design of large off-shore platforms. However, the difficulty that remains for offshore is the development of an effective shear source.

Acknowledgments

The assistance of the Natural Sciences and Engineering Research Council; The British Columbia Ministry of Highways and Transportation; Golder Associates, Vancouver; T. Lunne and T. Eidsmoen of the Norwegian Geotechnical Institute; L. Youd of the U.S. Geological Survey; J.L. Chameau, Purdue University and the technical staff of the Civil Engineering Department, University of British Columbia. The work of N. Laing and J. Greig is also appreciated.

References

1. ASTM, 1979, Designation: D3441, American Society for Testing and Materials, Standard Method for Deep Quasi-Static, Cone and Friction-Cone Penetration Tests of Soil.

2. Bennett, M.J., McLaughlin, P.V., Sarmiento, J.S. and Youd, T.L., 1984, "Geotechnical Investigation of Liquefaction Sites, Imperial Valley, California", U.S. Department of Interior, Geological Survey, Open-File Report No. 84-252.

3. Campanella, R.G. and Robertson, P.K., 1981, "Applied Cone Research", Sym. on Cone Penetration Testing and Experience, Geotechnical Eng. Div., ASCE, Oct., pp. 343-362.

4. Campanella, R.G., Robertson, P.K. and Gillespie, D., 1983, "Cone Penetration Testing in Deltaic Soils", Canadian Geotechnical Journal, Vol. 20, No. 1, pgs. 23-35.

5. Eidsmoen, T., Gillespie, D., Lunne, T. and Campanella, R.G., 1984, "Evaluation of the Seismic Cone Penetration Test", Norwegion Geotechnical Institute and University of British Columbia Joint Report.

6. Hoar, R.J. and Stokoe, K.H., 1978, "Generation and Measurement of Shear Waves In-situ", Dynamic Geotechnical Testing, ASTM STP 654, American Society for Testing Materials, pp. 3-29.

7. Nazarian, S. and Stokoe, K.H., 1984, "In-situ Shear Wave Velocities from Spectral Analysis of Surface Waves", Proceedings of the Eighth World Conference on Earthquake Engineering, Vol. III, pg. 31-38.

8. Robertson, P.K. and Campanella, R.G., 1983, "Interpretation of Cone Penetration Tests; Part I and II", Canadian Geotechnical Journal, Vol. 20, No. 4, pp. 718-745.

9. Seed, H.B. and Idris, I.M., 1970, "Soil Moduli and Damping Factors for Dynamic Response Analyses", Report to EERC 70-10, Earthquake Eng. Research Center, Univ. of California, Berkeley.

10. White, J.E., 1965, "Seismic Waves: Radiation, Transmission and Attenuation", McGraw Hill Publishing Co., New York.

SEISMIC ANALYSIS OF HORSE CREEK DAM, HUDSON, COLORADO

W.A. Charlie,M.,ASCE[1], G.T. Jirak[2] and D.O. Doehring[3]

INTRODUCTION

On April 8, 1981, six days after a Richter magnitude 4.0 earthquake, inspectors from the Colorado State Engineer's Office discovered a small crack in outboard side of Horse Creek Dam. Its presence prompted inspectors to request that the dam owners monitor the condition. Eight weeks after the earthquake, the State Engineer's Office was notified that the crest of the structure had settled approximately three feet and the outboard side of the dam was bulging along a zone 175 meters long. Horse Creek dam is located approximately 24 km northeast of Denver, Colorado near the town of Hudson. The structure is an homogeneous embankment 1,460 meters long and 17 meters high. The dam's cross section is shown in Figure 1. The grain size distribution of the silt/clay foundation and embankment material and the sand layer are given in Figure 2. The dam was built in 1911 for the Henerylyn Irrigation District and has a useful storage capacity of 18 million cubic meters (15,000 acre-feet). The following evidence favors the conclusion that Horse Creek Dam had experienced a dynamically-induced failure.

1. The presence of a low density sand layer in the substrate beneath the structure.

2. The presence of poorly compacted loess embankment material.

3. The temporal relationship to the earthquake.

4. The full reservoir conditions that existed at the time of the earthquake.

----

[1]Associate Professor, Department of Civil Engineering, Colorado State University, Fort Collins, Colorado.
[2]Geotechnical Engineer, Chen and Associates, Inc., Denver, Colorado.
[3]Associate Professor, Department of Earth Resources, Colorado State University, Fort Collins, Colorado.

5. The lack of interior drainage that produced a high phreatic surface within the structure.

6. The upstream and downstream slopes of the structure steeper than recommended for homogeneous dams (Bureau of Reclamation, 1977).

The following factors tend to discredit the conclusion of dynamically-induced failure.

1. A seismic event of Richter magnitude 4.0 is not generally thought to be harmful to dams.

2. The dam has experienced earthquakes of greater magnitude without failure.

3. Soils containing over ten percent clays are generally not sensitive to dynamic loading, (Seed, 1981).

4. Major distress conditions were not reported for 53 days following the earthquake.

The purpose of this study is to determine if the earthquake was a significant factor contributing to the distress of the Horse Creek Dam during the weeks that followed. In the course of our analysis, we shall also consider the likely effects of a 100-year design earthquake on this structure. Interpretation of the cause of failure is no simple matter as little is known about the behavior of saturated loess when subjected to earthquake loadings.

HISTORY OF HORSE CREEK DAM

Horse Creek Dam was constructed in 1911. Although the original plans are still available, most records prior to 1933 were destroyed in a fire. In general, design and construction of early embankments relied heavily on experience of the individual builder and careful material control was rarely employed.

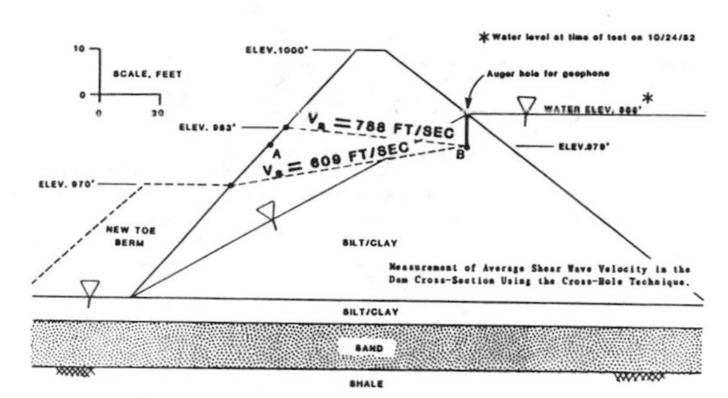

Measurement of Average Shear Wave Velocity in the Dam Cross-Section Using the Cross-Hole Technique.

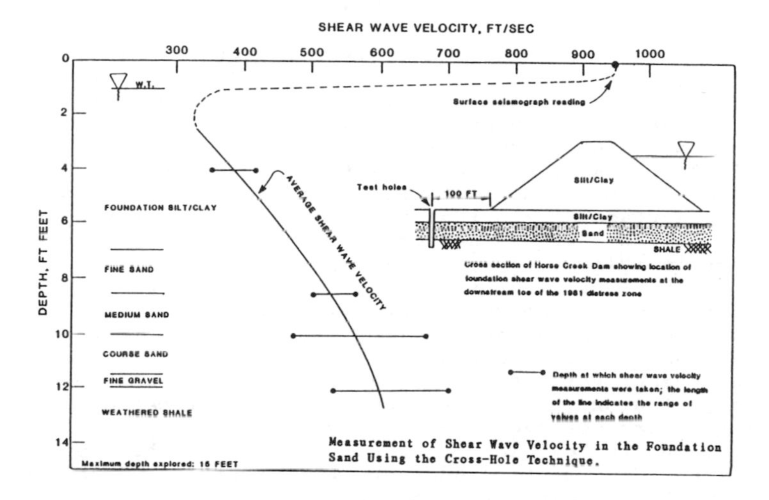

Measurement of Shear Wave Velocity in the Foundation Sand Using the Cross-Hole Technique.

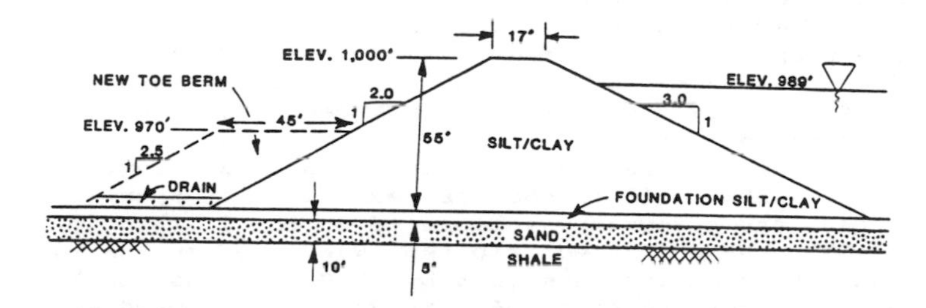

Figure 1. Cross Section of Horse Creek Dam.

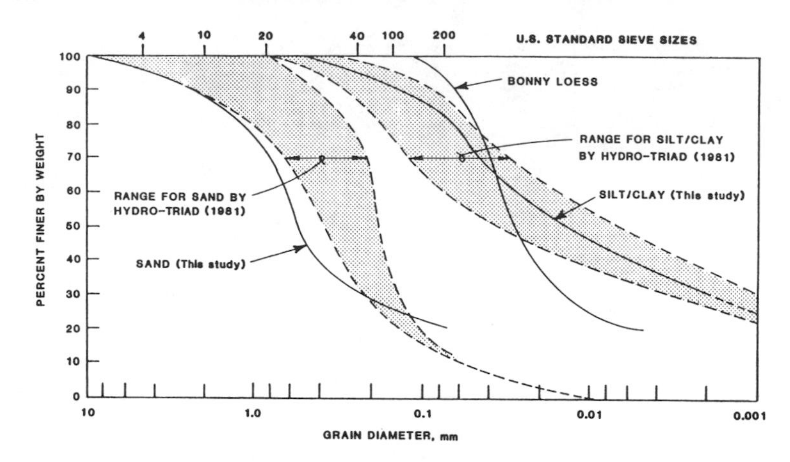

Figure 2. Grain Size Distribution of Soils.

Bedrock beneath the dam is Cretaceous shale and is overlain in most places by 5 to 6 meters of Quaternary sediments consisting of sand containing lenses of stiff clay overlain by silt-clay. Foundation preparation, at the time of construction, consisted of plowing the surface, removing the sod covering the upstream third of the site and placing it over the downstream section (Hinderlider, 1914).

The upstream face was lined with a 0.1 m thick barrier of reinforced concrete. Water was first diverted from the feeder canal to the reservoir in February, 1912 and allowed to rise to within 4.3 m of the crest. Settlement of as much as 2 m (13 percent of total height) along the crest and severe cracking of the liner were observed the first year.

On the evening of January 28, 1914, the dam failed along a 70 m section near the east abutment. The breach completely destroyed the concrete outlet works. The failure was attributed to piping, caused by insufficient compaction of the fill and to seepage along the outlet conduit. Due to the rapid lowering of the reservoir level, excess hydrostatic pressures caused the upstream concrete facing to slide some 3 to 6 meters down the very steep 1.5 to 1.0 slope (Hinderlider, 1914). Seismic activity was not recorded around the time of this failure.

The dam was rebuilt in 1926 (PRC Engineering, 1981) with a flattened upstream slope of approximately 3 to 1 and a downstream slope of 2 to 1 (Hinderlider, 1948). Hand placed rip-rap was used on the inboard face. Records indicate that by 1948 serious settlement of a large section of the dam had occurred and that the State Engineer's Office had requested that repairs be made. On several occasions during the years that followed 1948, fill was added to the crest to compensate for continued settlement. Several attempts have also been made to reduce seepage including the construction of a 175 meter long 3.5 meter deep bentonite-fill cutoff trench.

In 1973, a surficial slip failure occurred along a 45 meter section of the downstream side of the dam (Davis, 1973). The event was attributed to the combination of a high phreatic surface and the possible presence of localized porous material. The failure was not related to any known seismic events. Repair of the slide-damaged structure included the construction of a stabilizing berm along the east 300 meters of the downstream face of the dam. No significant damage to the dam was reported between 1971 and 1981.

EARTHQUAKE POTENTIAL IN COLORADO

Colorado is located in a region of low to moderate seismic activity, however, geologic and geophysical investigations have discovered several active faults that are capable of producing potentially damaging earthquakes. These investigations suggest that parts of Colorado have repeatedly experienced seismic events up to magnitude 7.5 during the past several million years (Kirkham and Rogers, 1981).

The first reported Colorado earthquake occurred in 1870 and during the following 110 years only a few have exceeded a Richter magnitude of 5.0.

Colorado is divided into six different seismotectonic provinces. The most severe earthquake for the region occurred on November 7, 1882 and is thought to have had a magnitude of 6.5 on the Richter Scale and an epicentral location north of Meeker, Colorado (McGuire et al., 1982). Current estimates of the maximum credible earthquake for that area are magnitude 6.5, the same as the 1882 event. The maximum credible earthquake is a postulated event whose magnitude is based on known active fault length, displacement and recent seismic history. The plains region of Colorado, where Horse Creek Dam is located, has only four known active faults and has an estimated maximum credible earthquake of Richter magnitude 5.5 to 6.0 (Kirkham and Rogers, 1981).

DYNAMIC ANALYSIS OF HORSE CREEK DAM

Horse Creek Dam was analyzed for liquefaction potential using Seed and Idriss' (1971) simplified approach, the threshold strain method of Dobry, et al., (1982) and Smart and Van Thun's (1983) empirical approach. Dynamic parameters of the soil were determined using laboratory and field tests as well as empirical methods (Jirak, 1983). Static and pseudo-static stability evaluations were performed by Hydro-Triad, Ltd., of Lakewood, Colorado utilizing measured phreatic levels and laboratory effective stress strength data. Soil samples used in the laboratory tests were taken from boreholes drilled at the centerline and downstream toe of the slide area. The static analysis, based on maximum strength parameters, resulted in a minimum factor of safety against slope instability of 1.68. The minimum factors of safety against slope instability for the two earthquake conditions considered, using the pseudo-static approach with horizontal coefficients of 0.02 g and 0.1 g, were 1.59 and 1.35 respectively (Hydro-Triad, Ltd., 1981). Both the static and pseudo-static methods of analysis fail to explain the slope movement.

The design earthquakes considered for the analysis include the magnitude 4.0 event of April 2, 1981 and the 100-year design earthquake. A third event, the maximum credible earthquake, was considered as an upper bound. A 100-year event was estimated as a magnitude 5.7 or greater based on the comprehensive record of seismic history given by Kirkham and Rogers (1981) and the seismic risk study performed for the Saint Vrain Nuclear Power Plant located approximately 60 km northwest of the dam (Public Service of Colorado Company, 1967). The maximum credible earthquake was estimated as magnitude 6.2 based on the 16 km length of the Derby Fault. Seismic risk from other, more distant, active faults was also considered but do not exceed the risk from the Derby Fault. The epicentral locations and the magnitudes of earthquakes associated with the Derby Fault are shown in Figure 3.

The April 2, 1981 earthquake epicenter, shown in Figure 3, was located approximately 30 km from the dam and had a magnitude of 4.0. Maximum rock acceleration beneath the dam was probably less than 0.02 g. To be consistent with the Hydro-Triad analysis and to allow for some scatter in the data, 0.02 g was assumed as the maximum rock acceleration at the dam. The predominant period was estimated as less than 0.2 seconds (greater than 5 Hz) and the duration as less than five seconds. The maximum rock acceleration at Horse Creek Dam due to the 100-year event on the Derby Fault was estimated at 0.16 g, the predominate period is 0.24 seconds (4.2 Hz) and the duration of strong motion (i.e., acceleration greater than 0.05 g) is eight seconds (Horn and Scott, 1977). The April 2, 1981 and the postulated 100-year earthquake were simulated by digitized accelerograms of the Pasadena recording of the 1952 Kern County, California earthquake (Schnabel et al., 1972; Seed, 1967). The Pasadena recording has a maximum acceleration of 0.0579 g, a predominant period of 0.65 seconds (1.5 Hz) and a Richter magnitude of 7.7. The accelerograms used in the response analysis for Horse Creek Dam were adjusted by scaling the digitized accelerations of the Pasadena recording and the predominant period was modified by adjusting the time step between digitized inputs. Based on the approach given by Sarma (1979) the natural period of the dam is about 0.3 seconds at shear strains less than about 10^{-3} percent. As strains increase during a seismic event, the shear modulus decreases and the predicted natural period increases to about 0.5 seconds at 10^{-1} percent shear strains.

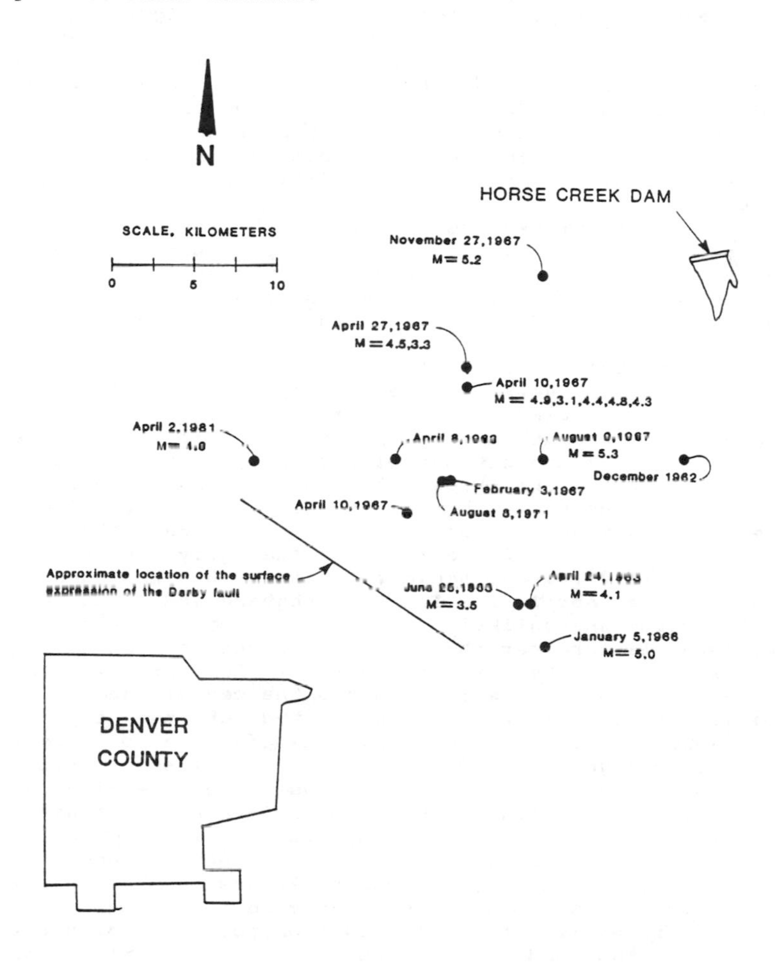

Figure 3. Seismicity Map of the Region of Derby Fault and Horse Creek Dam (after Hydro-Triad, 1981).

The Hardin-Black equations were used to obtain the maximum shear modulus at each location. These equations were verified by using cross-hole seismic techniques. The curves given by Seed (1976) were used to adjust the shear modulus and damping for shear strains.

Cyclic triaxial test procedures were used to determine the cyclic strength of the soil at Horse Creek Dam. Dynamic failure curves were developed for the embankment silt/clay and for the foundation sand. The results of these tests are shown in Figures 4 and 5. A representative sample of the embankment material was taken from a 2 meter deep hand augered hole located on the downstream face of the dam. A 4.5 meter deep, hand augered hole located downstream from the toe of the dam was used to sample the foundation sand. Reconstituted samples were compacted at near field density and at estimated initial field placement water content.

The calculated safety against liquefaction is assessed by comparing the soil's cyclic strength with the earthquake induced stresses. If the cyclic strength for an equivalent number of cycles is less than the earthquake induced stress, then initial liquefaction will occur. The irregular stress history, as developed by the computer model for each element, is converted to an equivalent uniform stress magnitude by multiplying the peak irregular stress times a reduction factor. Seed (1976) recommends a reduction factor for the peak stress of 0.65 and has correlated the equivalent number of cycles of uniform stress to the earthquake magnitude. For both earthquake conditions at Horse Creek Dam (the 4.0 magnitude of April 2, 1981 and the 5.7 magnitude for the 100-year event) the equivalent number of cycles is five. As shown in Figure 6, the analysis of the April 2, 1981 earthquake indicates that liquefaction was unlikely since the dynamic strength envelope is greater than the equivalent uniform stress that was induced by the earthquake. Furthermore, the minimum factor of safety along the centerline of the dam is estimated at 2.2 and is located at the top of the sand layer. Typical minimum factors of safety used for design are 1.3 to 1.5 (Khilnani, 1982), therefore, the dam should be considered safe with respect to liquefaction. For the 100-year earthquake, the minimum factor of safety against liquefaction is equal to 1.2. This is below the accepted range of safety factors, but may be acceptable due to the conservative assumptions used in this analysis.

To summarize the liquefaction potential of Horse Creek Dam, the sand layer beneath the centerline of the dam is most susceptible to liquefaction. However,

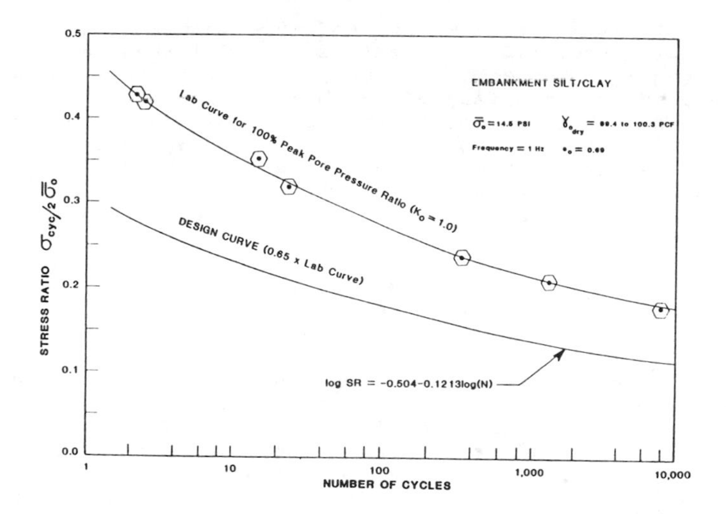

Figure 4. Cyclic Triaxial Dynamic Failure Curve for the Silt/Clay.

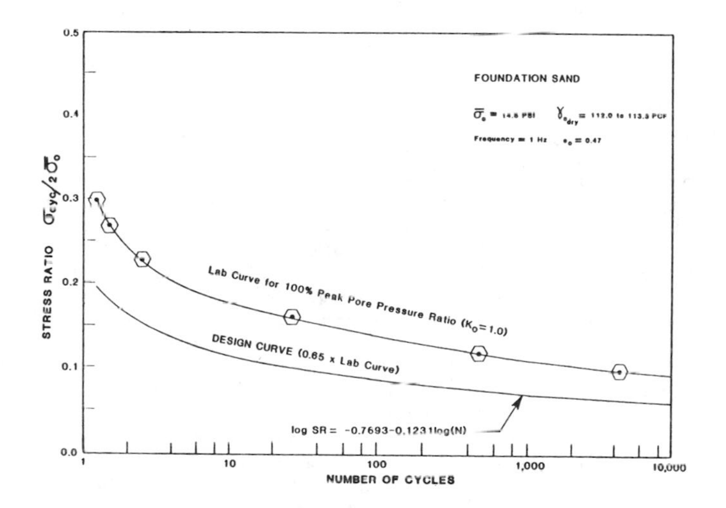

Figure 5. Cyclic Triaxial Dynamic Failure Curve for the Foundation Sand.

the minimum factors of safety against liquefaction of 2.2 and 1.2 for the April 2, 1981 and the 100-year earthquakes respectively are not sufficiently low to jeopardize the safety of the dam.

An alternate method to the stress controlled cyclic triaxial test is the strain controlled approach. Cyclic strains below the threshold will not develop excess residual pore pressures. Dobry, et al. (1982) have shown that the threshold strain for granular soils is in the range of 1×10^{-2} to 3×10^{-2} percent. Above the threshold, pore pressures increase as the magnitude of the strain increases. Although laboratory tests for Horse Creek Dam were stress controlled, the low stress levels being investigated allow for a reasonably accurate assessment of the threshold strains. Our threshold strains were 1×10^{-2} percent for the foundation sand and 3×10^{-2} percent for the embankment material.

The values of threshold strain and the computed equivalent uniform strains are plotted with depth along the centerline of the dam in Figure 7. The induced strain from the April 2, 1981 earthquake does not exceed the threshold strain along most of the centerline of the dam. However, the foundation sand may experience some pore pressure increase. The induced strains from the 100-year design earthquake however, exceed the threshold strain everywhere along the centerline, thus indicating increases in pore pressures.

Using the standard penetration blow counts obtained by Hydro-Triad (1981) and the empirical relations given by Seed (1967), an estimate of the soil's dynamic strength can be determined. The uniform cyclic shear stress required to cause liquefaction and the computed equivalent cyclic shear stress are plotted in Figure 8. For the April 2, 1981 earthquake, the minimum factor of safety against liquefaction is 2.35 in the sand layer. This value compares with 2.2 calculated from the procedure using the laboratory test data. For the 100-year design earthquake, the minimum factor of safety is in the sand layer and equals 1.1, which compares with the 1.2 obtained from the procedures discussed above. Using an average shear wave velocity of 200 to 250 m/sec, the stress ratios necessary to cause initial liquefaction, fall in the same range of values as those obtained by using modified blow counts. Epicentral distances from Horse Creek Dam versus magnitude are shown on Figure 9. Data plotting below the line shown on the graph are not expected to experience liquefaction (Smart and Von Thun, 1983). The April 2, 1981 earthquake and the 100-year design earthquake plot below the line in Figure 9. This is consistent with the results of the empirical and laboratory analysis discussed above.

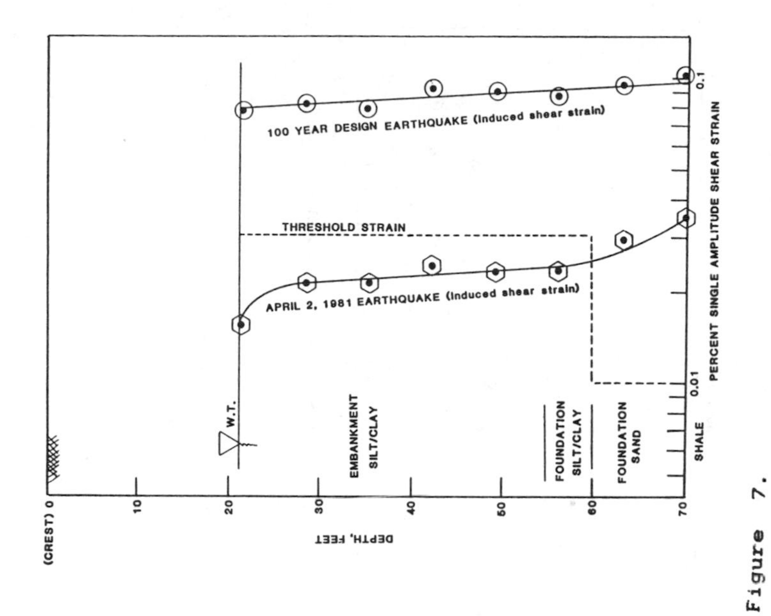

Figure 7.
Earthquake Induced Strains Versus Threshold Strains.

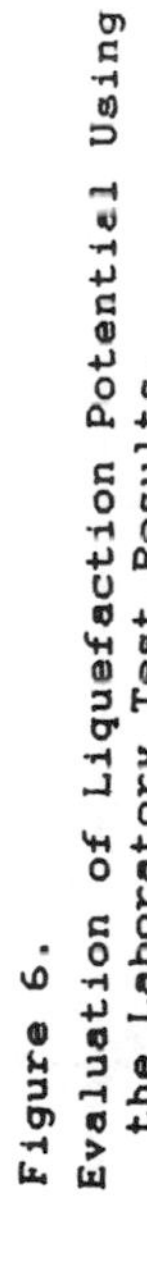

Figure 6.
Evaluation of Liquefaction Potential Using the Laboratory Test Results.

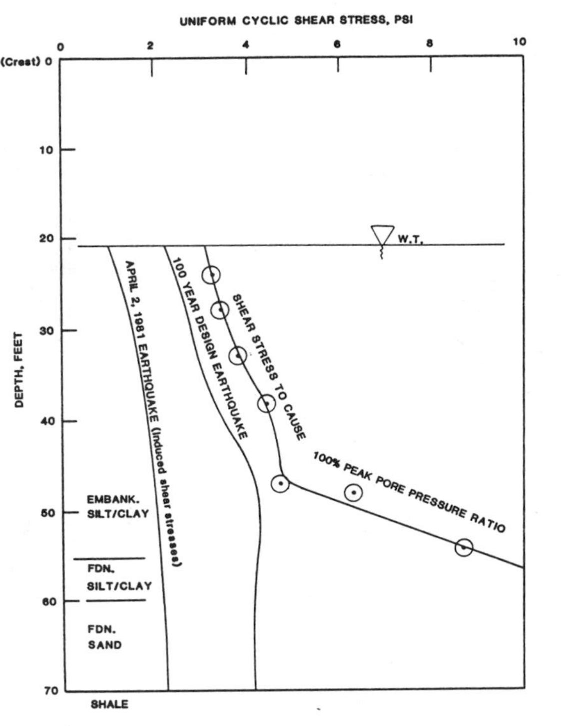

Figure 8.
Evaluation of Liquefaction Potential Using Standard Penetration Blow Counts.

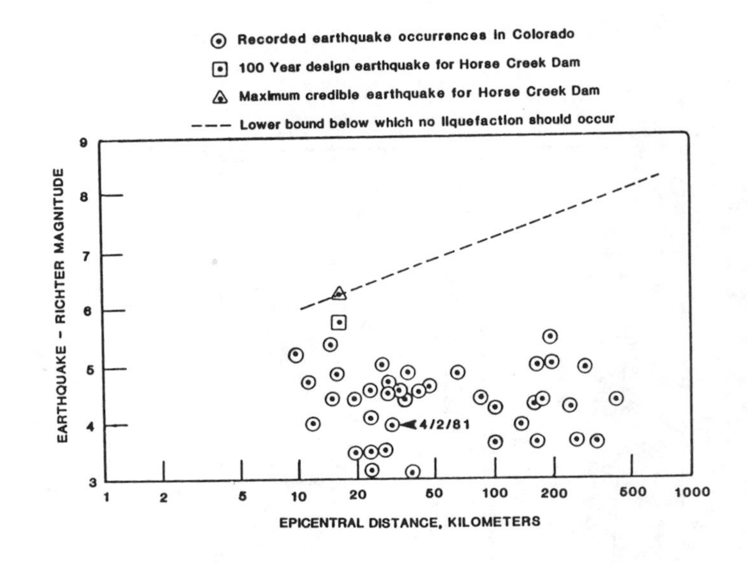

Figure 9.
Liquefaction Potential Versus the Bounds for Liquefaction Elsewhere (Smart and Von Thun, 1983).

CONCLUSIONS

It is reasonable to conclude that the magnitude 4.0 earthquake of April 2, 1981 was not a significant fact contributing to the failure at Horse Creek Dam. The increased pore pressure resulting from the earthquake were calculated to be less than four percent of the initial effective overburden stress at the centerline of the dam. These increases in pore pressure were not of sufficient magnitude to cause significant reductions in strength and, hence, stability of the structure. The analysis of the 100-year design earthquake results in a minimum calculated factor of safety against liquefaction of 1.2. The excess pore pressure generated by the earthquake is about ten percent of the effective overburden stress at the centerline of the dam. With such a marginal factor of safety, local zones of liquefaction could occur. Since the 1981 failure, a berm has been added to the toe of the dam on the downstream side.

REFERENCES CITED

Bureau of Reclamation, 1977, Design of Small Dams, A Water Resources Technical Publication: U.S. Dept. Interior, Government Printing Office, Washington, D.C.

Davis, F.J., 1973, Report on Horse Creek Dam: unpublished report, Colorado State Engineer's Office.

Dobry, R., R.S. Ladd, F.Y. Yokel, R.M. Chung and D. Powell, 1982, Prediction of Pore Pressure Buildup and Liquefaction of Sands During Earthquakes by Cyclic Strain Method: Nat. Bur. Standards, Building Series 138, U.S. Dept. Commerce.

Hinderlider, M.C., 1914, Failure of Horse Creek Earth Dam: Engineering News, Vol. 71, No. 16.

________ , 1948, unpublished letter to R.D. Culverwell, Secretary, Henerylyn Irrigation District.

Horn, B.,and M. Scott, 1977, Geological Hazards: Second Edition, Springer-Verlag.

Hydro-Triad, Ltd., 1981, Horse Creek Dam Geotechnical Investigations and Design of Remedial Repairs: Consulting Report, Lakewood, Colorado.

Jirak, G.T., 1983, Seismic Analysis of Horse Creek Dam, Hudson, Colorado: M.S. Thesis, Colorado State University, Fort Collins, Colorado.

Khilnani,K.S., 1982, Seismic Stability of the Revelstoke Earthfill Dam: Canadian Geotech. Jour., Vol. 19, p. 63-75.

Kirkham, R.M. and M.P. Rogers. 1981, Earthquake Potential in Colorado: Bull. 43, Colo. Geol. Survey, Denver, Colorado.

McGuire, R.K., A. Krusi and S.D. Oaks, 1982, The Colorado Earthquake of November 7, 1882: Size, Epicentral Location, Intensities and Possible Causative Fault: The Mountain Geologist, Vol. 19, No. 1, p. 11-23.

Public Service of Colorado Company, 1967, Fort Saint Vrain Nuclear Generating Station: Final Safety Analysis and Report: Denver, Colorado.

Sarma, F.K., 1979, Response and Stability of Earth Dams During Strong Earthquakes: Misc. Paper GL-79-13, U.S. Waterway Exper. Station, U.S. Army, Corps of Engineers, Vicksberg, Miss.

Schnabel, P.B., J. Lysmer and H.B. Seed, 1972, SHAKE: A Computer Program for Earthquake Response Analysis of Horizontally Layered Sites: Report EERC 72-12, Earthquake Engin. Res, Center, Univ. Calif. Berkeley, California.

Seed, H.B., 1967, Turnigan Heights: Jour. Soil Mech. and Found. Engin. Div., A.S.C.E., SM4.

Seed, H.B. and I.M. Idriss, 1971, Simplified Procedure for Evaluating Liquefaction Potential: Jour. Soil Mech. and Found. Engin. Div., A.S.C.E., SM9.

________ , 1981, Earthquake-Resistant Design of Earth Dams: Proc., Intl. Conf. on Recent Advances in Geotech. Earthquake Engin. and Soil Dynamics, Vol. II, Univ. Missouri, Rolla, Mo.

Smart, J.D. and J.L. Von Thun, 1983, Seismic Design and Analysis of Embankment Dams: Recent Bureau of Reclamation Experience: Proc., A.S.C.E. Mtg., Philadelphia, Penn.

RELATIONSHIPS BETWEEN SHEAR WAVE VELOCITY AND DEPTH OF OVERBURDEN

by
Marshall Lew, M. ASCE (1)
and
Kenneth W. Campbell, M. ASCE (2)

ABSTRACT

Relationships between in-situ shear wave velocity and depth of overburden have been derived for various Quaternary age soils encountered primarily in Southern California. The geotechnical parameters found to be significant were geologic age, gravel content, water table depth, and dry density, as well as depth of overburden. The data base for deriving the relationships consists of over 270 shear wave velocity surveys which included refraction, downhole, and cross-hole surveys. The surveys generally provided data to depths of about 100 feet (about 30 meters), however, several of the surveys provided shear wave velocity data to depths of 200 feet (about 60 meters) and greater. Compressional wave velocity data was obtained in many of the same surveys and relationships between compression wave velocity and depth of overburden are also presented.

INTRODUCTION

This paper presents relationships among seismic velocity, depth, and various geotechnical parameters, the most important of which, is depth of overburden. These correlations are based on a geotechnical classification system that defines ranges of seismic velocity according to soil or rock type, geologic age, gravel content, water table depth, dry density, and depth of overburden.

Such relationships were first introduced by Campbell and Duke (1976) relating shear-wave velocities and various geotechnical characteristics. These relationships were based on seismic velocity measurements at 63 sites conducted in the greater Los Angeles area after the 1971 San Fernando earthquake (see Duke, et al, 1971 and Eguchi, et al, 1976). The first study was updated by incorporating velocity measurements made at 48 other sites (Campbell, et al, 1979 and Lew, et al, 1981). This current study supplements the prior works with measurements

---

(1) Vice President, LeRoy Crandall and Associates, (a Subsidiary of Law Engineering Testing Company), 711 North Alvarado Street, Los Angeles, California 90026-4099

(2) Research Civil Engineer, United States Geological Survey, MS 966, Box 25046, Federal Center, Denver, Colorado 80225-0046

made at 160 additional sites. Most of the supplemental measurements, like the prior measurements, were obtained at sites in Southern California, although there were a few from Northern California. Most of these data were obtained from borings drilled as a part of a geotechnical investigation. These new data were used to establish velocity relationships, and to improve and extend the geotechnical classification system first developed by Campbell and Duke.

MEASUREMENT TECHNIQUES

The large majority of the measurements were performed with downhole surveys. The remainder were performed with surface refraction or crosshole surveys. Detailed descriptions of these techniques, together with their strengths and weaknesses, have been discussed extensively in the literature and will not be repeated here. The predominant soil type for each layer was identified from exploration borings at the site of each survey. In most cases, this information was available from the boring in which the survey was conducted in the case of downhole and crosshole surveys. Soils were classified using the Uniform Soil Classification System (1960). The approximate quantity of gravel, debris, and organic content were determined. Materials classified as hydraulic fill, engineered fill, or non-engineered fill represent man-made deposits. The geology of each site was obtained from a geologic investigation which was usually part of the overall geotechnical investigation.

Water levels were established from exploration borings at the site when possible. Where such data were not available or where the water levels were below boring depths, nearby water well data were used to establish the depths.

The shear-wave and compression-wave velocities determined were the average velocities over the depth ranges. At least two points within the layer were used to established these velocities. Velocities were not generally computed for layers of less than about five feet (1.5 meters) in thickness. Exceptions to this were thin surface layers and low velocity layers at depth where the geophone spacing was reduced accordingly.

Poisson's ratios (ν) were computed where the P-wave velocity (V_p) and the S-wave velocity (V_s) were determined by the expression (Richart, et al, 1970):

$$\nu = 1/2 \; \frac{(V_p/V_s)^2 - 2}{(V_p/V_s)^2 - 1} \qquad (1)$$

Equation (1) was developed assuming that near surface soils can be treated as an elastic medium for the low dynamic shear strains (on the order of 10^{-4}% or less) induced in shallow seismic surveys.

GEOTECHNICAL CLASSIFICATION SYSTEM

The Geotechnical Classification System developed by Campbell, et al (1979) was used to classify the soil and rock types of each layer encountered in the numerous surveys. The components of this system are diagrammed in Figure 1, Geotechnical Classification System. Table 1 gives a detailed description of the classification system which includes the effects of water table, gravel content, dry density, soil firmness, rock hardness, and soil or rock type.

VELOCITY CORRELATIONS

Statistical studies of seismic velocity were performed with the velocity and geotechnical data collected for the various surveys. Deposits were classified in accordance with the Geotechnical Classification System presented in the previous section. In this paper, results are shown for natural soils, and in particular, soils with gravel content less than 10 percent. Results for the other categories of soil and rock will be presented sometime in the future.

NEAR SURFACE VELOCITIES

Statistical analyses of the surface deposits and a few near surface deposits were performed on both S-wave and P-wave velocities. Table 2 shows the results of the analyses; the mean and the standard deviations of the shear wave and compressional wave velocities are presented.

RELATIONSHIP BETWEEN SHEAR-WAVE VELOCITY AND DEPTH

It has been found in the previous study (Campbell, et al, 1979) that the relationship of shear-wave velocity with depth could be adequately represented by the following expression:

$$V_o = K(d + c)^n \qquad (2)$$

where V_s is shear-wave velocity, d is depth, and K, c and n are constants dependent upon the geotechnical classification. Term c accounts for non-zero values of V_s at the ground surface. Thus, as d approaches zero, shear-wave velocity takes on the value $Vs = Kc^n$.

In order that a regression analysis could be used to estimate the constants K, d, and n for each of the geotechnical classifications, Equation (2) was linearized by taking the natural logarithm of both sides, giving

$$\text{Ln } V_s = \text{Ln } K + n \text{ Ln } (d + c) \qquad (3)$$

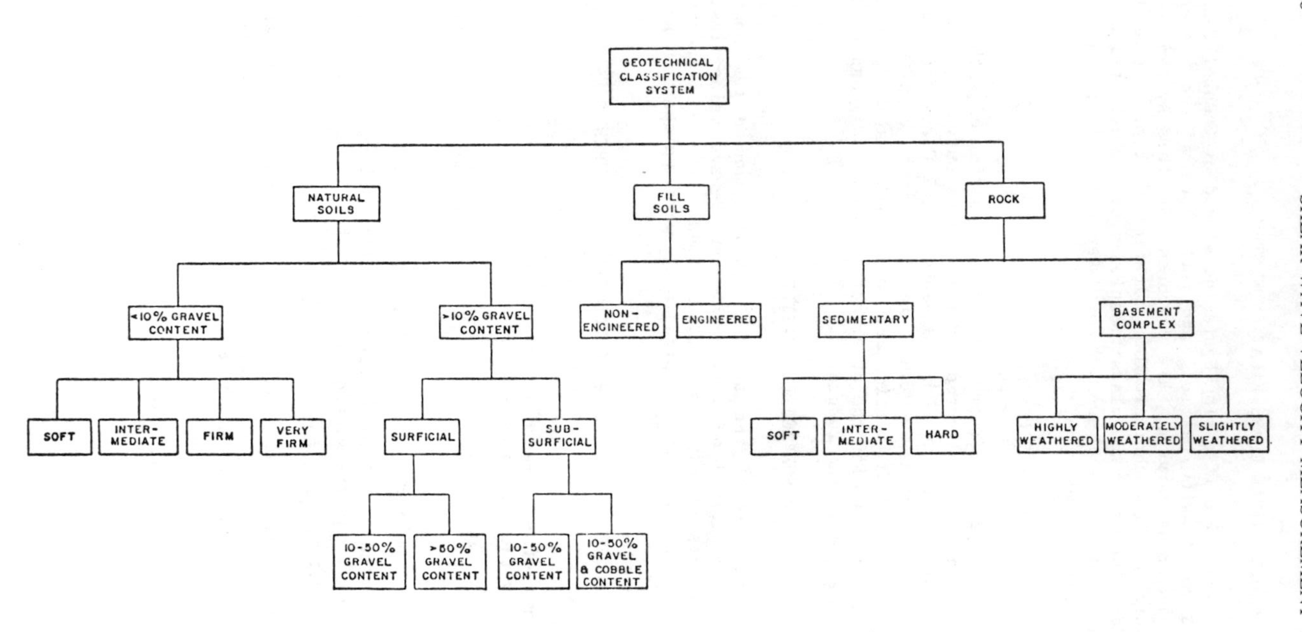

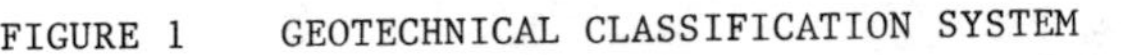
FIGURE 1 GEOTECHNICAL CLASSIFICATION SYSTEM

TABLE 1

GEOTECHNICAL CLASSIFICATION SYSTEM

Geotechnical Classification			Description	Dry Density (Lbs./Cu.Ft.)	Depth (Ft.)	Geologic Age and Unit
Natural Soils	<10% Gravel and Cobbles	Soft	Very recent floodplain, lake, swamp, estuarine, and delta deposits, and hydraulic fill soils; may contain organics.	<100	≥ 0	Holocene
			Saturated Holocene age soils.	90-110	>10-15	Holocene
			Very low density soils; primarily fine-grained.	<90	≥ 0	Holocene
			Soft to moderately soft clays and clayey silts; generally dark grey to black.	<100 <80	0 ≥ 10	Holocene and Pleistocene Holocene and Pleistocene
		Intermediate	Unsaturated Holocene age soils of moderate density; may contain moderate amounts of gravel at depths below 30 to 50 feet.	90-110	≥ 0	Holocene
			Saturated, uncemented Pleistocene and Eocene age soils; may contain moderate amounts of gravel at depths below 10 to 20 feet.	95-115	>10-15	Pleistocene and Eocene
			Low density Pleistocene and Eocene age soils.	90-100	≥ 0	Pleistocene and Eocene
		Firm	High density Holocene age soils.	>110	≥ 0	Holocene
			Unsaturated, uncemented Pleistocene and Eocene age soils of moderate density; may contain moderate amounts of gravel at depths below 10 to 20 feet	100-115	≥ 0	Pleistocene and Eocene
		Very Firm	Pleistocene and Eocene age soils of high density.	>117	≥ 0	Pleistocene and Eocene
	>10% Gravel and Cobbles	Surficial	Natural soils and engineered fill soils occurring at the surface containing gravel and cobbles or bounders; predominantly coarse grained:			
			10-50% gravel; some cobbles and boulders.	115-130	0	Holocene, Pleistocene and Eocene
			>50% gravel; some cobbles and boulders.	125-135	0	Holocene, Pleistocene, and Eocene
		Subsurficial	Natural soils occurring at depth containing gravel and cobbles or boulders; predominantly coarse grained:			
			10-50% gravel; some cobbles and boulders.	115-130 115-130	5-30 5-10	Holocene Pleistocene and Eocene
			10-50% gravel, cobbles, and boulders.	125-135 125-135	5-50 5-20	Holocene Pleistocene and Eocene

TABLE 1
(Continued)

Geotechnical Classification			Description	Dry Density (Lbs./Cu.Ft.)	Depth (Ft.)	Geologic Age and Unit
Fill Soils	Non-Engineered	---	Loose or slightly compacted man-made fill soils (excluding hydraulic fill); containing <10% gravel and cobbles.	100-115	0	---
	Engineered	---	Mechanically compacted man-made fill and natural soils; containing <10% gravel and cobbles.	110-125	0	---
Rock	Sedimentary	Soft	Highly weathered, low density siltstones and shales; usually light brown to light grey; Miocene shales are highly siliceous and diatomaceous.	65-90	15-50	Pliocene and Miocene
		Intermediate	Moderately weathered, moderate density siltstones and shales; Pliocene siltstones are primarily massive and dark grey. Miocene shales are highly siliceous and diatomaceous.	90-105	≥ 0	Pliocene and Miocene
		Hard	Moderately weathered, moderate to high density Miocene siltstones, shales, sandstones, and conglomerates.	>95	≥ 0	Miocene
	Basement Complex	Highly Weathered	Igneous and Metamorphic rock; highly weathered.	---	≥ 0	Mesozoic
		Moderately Weathered	Moderately weathered and fractured Igneous and Metamorphic rock.	---	≥ 0	Mesozoic
		Slightly Weathered	Unweathered or slightly weathered and fractured Igenous and Metamorphic rock.	---	≥ 0	Mesozoic

TABLE 2
SUMMARY OF NEAR SURFACE VELOCITIES

SOIL DESCRIPTION	SHEAR WAVE VELOCITIES (ft/sec)		COMPRESSIONAL WAVE VELOCITIES (ft/sec)	
	Mean	Standard Deviation	Mean	Standard Deviation
Soft Natural Soil	528	58	1169	320
Soft Clay (depth <10')	310	87	1090	--
Soft Clay (depth = 10-100')	630	69	--	--
Intermediate Natural Soil	701	132	1418	363
Firm Natural Soil	873	152	1672	400
Non-Engineered Fill	518	56	1324	244
Engineered Fill	867	--	1600	--
10 - 50% Gravel (depth = 0)	1040	--	2250	--
10 - 50% Gravel (depth = 5-60')	1305	188	2453	234
10 - 50% Gravel with Cobbles, 50% Gravel (depth = 5-50')	1599	409	3245	728
Saturated Soil	--	--	5265	435

Note: 1 ft/sec = 0.3048 m/sec

A summary of the regression analysis for soft natural soils, intermediate and saturated firm natural soils, and firm natural soils is given in Table 3; shear-wave velocity is expressed in units of feet per second and depth in units of feet. The depth was taken as the vertical distance from the ground surface to the top of the layer, not to the midpoint of the layer as some investigators have used. For the surface layer, the depth was taken as one-third the thickness of the layer.

TABLE 3
SUMMARY OF REGRESSION ANALYSIS
FOR SHEAR WAVE VELOCITY IN FEET/SECOND

$$\text{Ln } V_s = \text{Ln } K + n \text{ Ln } (d+c)$$

SOIL CLASSIFICATION	REGRESSION COEFFICIENTS Ln K	n	c	STANDARD DEVIATION
Soft Natural Soils	5.393	0.385	5.33	0.192
Intermediate and Saturated Firm Natural Soils	5.568	0.402	5.24	0.148
Firm Soils	6.259	0.280	0.54	0.145

The relationships found between the S-wave velocities with depth are shown in Figures 2, 3, and 4. Deep downhole seismic velocity data from El Centro and Cholame obtained in a study by Shannon & Wilson and Agbabian Associates (1975) was also included in the analyses of the soft natural soils category.

Because there were significant compressional wave velocity measurements, similar regression analysis was performed on the compressional wave velocities using an equation similar in form to equation (3):

$$\text{Ln } V_p = \text{Ln } K + n \text{ Ln } (d+c) \tag{4}$$

The results of these studies are presented in Table 4. In addition, a regression was performed on the combined shear and compressional wave data; the linearized form of the unified velocity equation for both shear and compressional wave velocity in units of feet per second is given by:

$$\text{Ln } V_{s,p} = \text{Ln } K = n \text{ Ln } (d+c) + m P \tag{5}$$

where P is equal to unity for P-waves and equal to zero for S-waves, and m is a determined constant for each geotechnical classification. The results of this analysis is presented in Table 5.

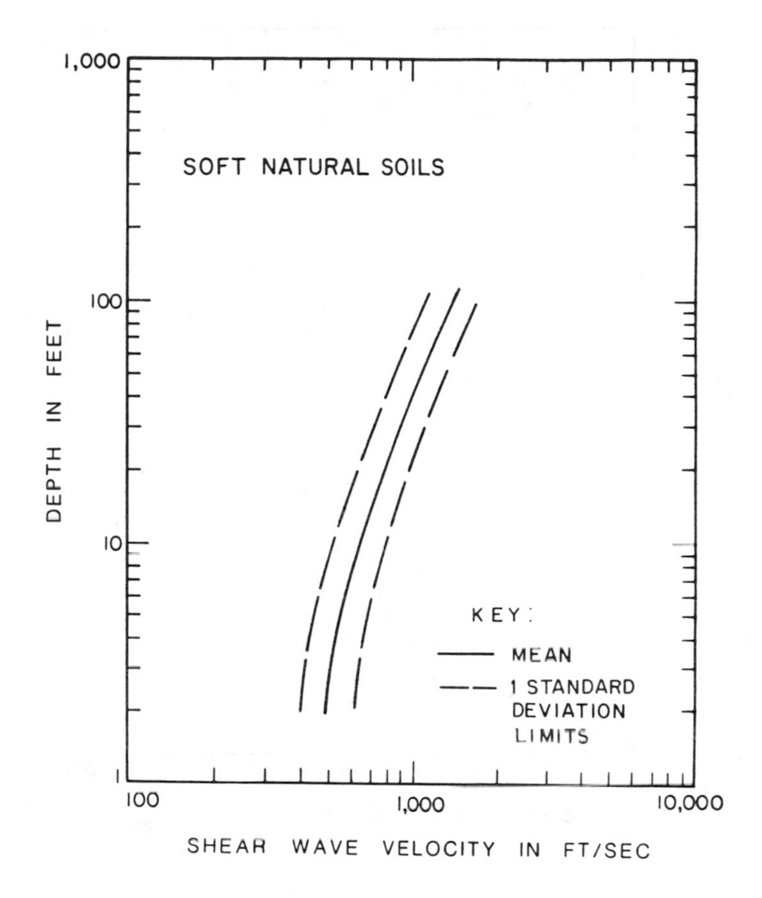

FIGURE 2
SHEAR WAVE VELOCITY VERSUS DEPTH FOR SOFT NATURAL SOILS
(Note: 1 ft = 0.3048 m and 1 ft/sec = 0.348 m/sec)

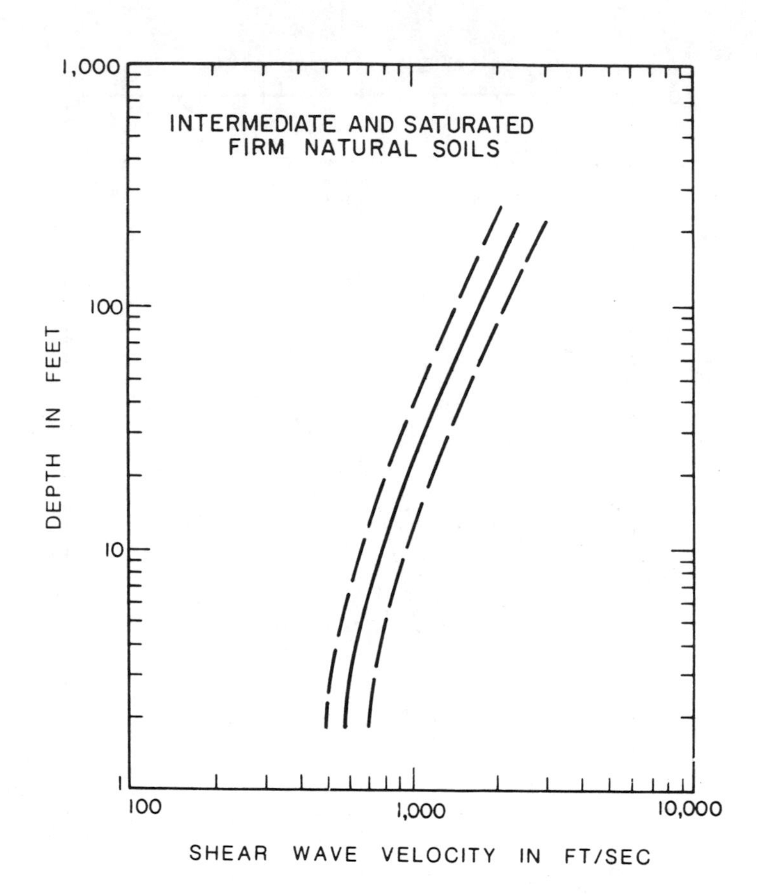

FIGURE 3
SHEAR WAVE VELOCITY VERSUS DEPTH FOR INTERMEDIATE AND SATURATED NATURAL SOILS
(Note: 1 ft = 0.3048 m and 1 ft/sec = 0.3048 m/sec)

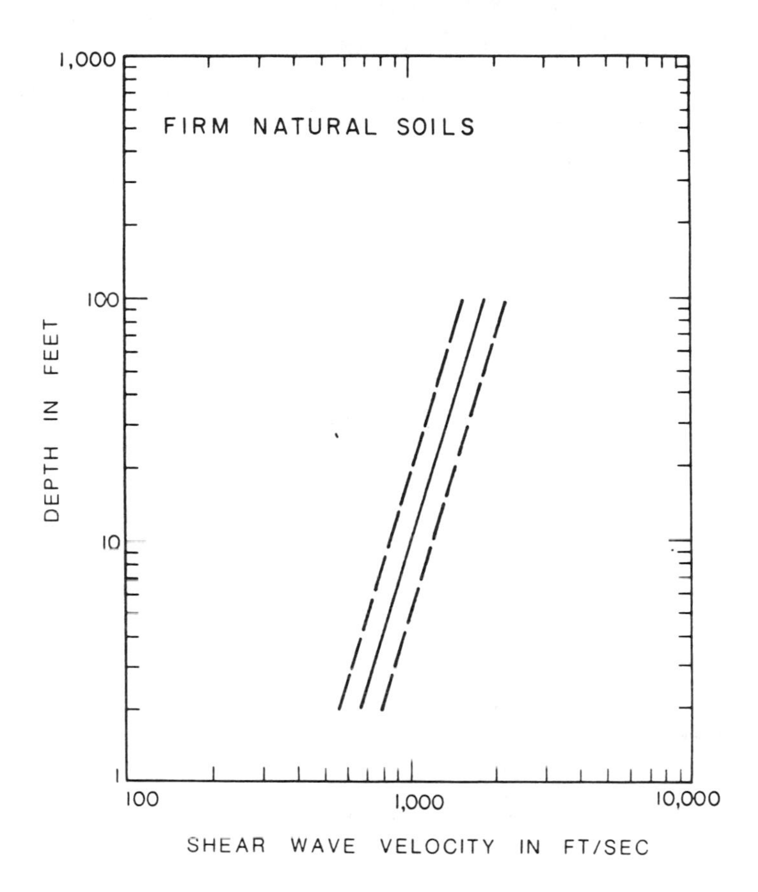

FIGURE 4
SHEAR WAVE VELOCITY VERSUS DEPTH FOR FIRM NATURAL SOILS
(Note: 1 ft = 0.3048 m and 1 ft/sec = 0.3048 m/sec)

TABLE 4
SUMMARY OF REGRESSION ANALYSIS
FOR COMPRESSIONAL WAVE VELOCITY IN FEET/SECOND

$$\text{Ln } V_p = \text{Ln } K + n \text{ Ln } (d+c)$$

SOIL CLASSIFICATION	REGRESSION COEFFICIENTS Ln K	n	c	STANDARD DEVIATION
Soft Natural Soils	(NOT SUFFICIENT DATA)			
Intermediate and Saturated Firm Natural Soils	6.622	0.305	1.40	0.178
Firm Soils	6.244	0.288	0.94	0.192

TABLE 5
SUMMARY OF REGRESSION ANALYSIS FOR
SHEAR AND COMPRESSIONAL WAVE VELOCITY IN FEET/SECOND

$$\text{Ln } V_{p,s} = \text{Ln } K + n \text{ Ln } (d+c) + m P$$

SOIL CLASSIFICATION	REGRESSION COEFFICIENTS Ln K	n	c	m	STANDARD DEVIATION
Soft Natural Soils	5.437	0.375	4.75	0.680	0.201
Intermediate and Saturated Firm Natural Soils	5.698	0.370	3.89	0.675	0.160
Firm Natural Soils	6.244	0.284	0.73	0.649	0.169

Note: P = 1 for P-waves, 0 for S-waves.

POISSON'S RATIO

Poisson's ratio was estimated from the relationships found in this study. The shear and compressional wave velocities determined from Equations (3) and (4) using the regression analysis determined constants were used in Equation (1) to compute Poisson's ratio for the three soil classifications. Poisson's ratio was also computed by using the shear and compressional wave velocities determined by Equation (5). The results are summarized in Table 6.

TABLE 6
COMPUTED VALUES OF POISSON'S RATIO

SOIL CLASSIFICATION	POISSON'S RATIO Using Equations (3) & (4)	POISSON'S RATIO Using Equation (5)
Soft Natural Soils	--	0.335
Intermediate and Saturated Firm Natural Soils	0.317	0.325
Firm Natural Soils	0.314	0.312

It is interesting to note that there is little difference in the value of Poisson's ratio for the three soil classifications.

CONCLUSIONS

Statistically derived relationships between shear and compressional wave velocities have been derived for several categories of Quaternary age soils. These relationships may provide a guide for estimating the seismic velocities from a limited amount of geotechnical data. However, it should be noted again that the velocities are based on measurements made at low shear strain levels less than 10^{-4}% and some reductions would be needed to be applied to account for strain softening at higher shear strain levels.

Several observations about the relationships between shear wave velocity and depth of overburden need be made. In making comparison between the present study and the prior study by Campbell, et al (1979) in which the same method of analysis was used, it should be noted that differences were observed. For the soft natural soils, it was observed that the current relationships were similar to those of the previous study, but the velocities near the surface were found to be higher. It was also observed that the current relationship for firm natural soils gives lower velocities than the previous relationship. One possible explanation for this observation could be that a large part of the new measurements were made in a geologic formation which is late Pleistocene in age. The earlier study had a larger proportion of data from older Pleistocene materials.

These relationships should be treated as preliminary. Further scrutiny of the firm natural soils may be needed and subdivision of the data into additional categories may be required. As the vast majority of the data was obtained in Southern California, caution should be utilized when applied to other geographic locations.

ACKNOWLEDGEMENTS

The authors would like to thank LeRoy Crandall and Associates for making available the additional data for use in this study. The authors would also like to thank Pamela Grizz, Todd Tostado, and Magda Ghika for their help in preparing this paper.

REFERENCES

Campbell, K.W., R. Chieruzzi, C.M. Duke, and M. Lew, "Correlations of Seismic Velocity in Depth in Southern California", School of Engineering and Applied Science, Report No. UCLA-ENG-7965, University of California, Los Angeles, October 1979.

Campbell, K.W. and C.M. Duke, "Correlations Among Seismic Velocity, Depth and Geology in the Los Angeles Area", School of Engineering and Applied Science Report No. UCLA-ENG-7662, University of California, Los Angeles, June 1976.

Corps of Engineers, "The Unified Soil Classification System", U.S. Army Technical Memorandum No. 3-357, Vol. 1, March 1953 (Revised April 1960).

Duke, C.M., J.A. Johnson, Y. Kharraz, K.W. Campbell, and N.A. Malpiede, "Subsurface Site Conditions and Geology in the San Fernando Earthquake Area", School of Engineering and Applied Science Report No. UCLA-ENG-7653, University of California, Los Angeles, June 1976.

Eguchi, R.T., K.W. Campbell, C.M. Duke, A.W. Chow, and J. Paternina, "Shear Velocities and Near-Surface Geologies at Accelerograph Sites that Recorded the San Fernando Earthquake", School of Engineering and Applied Science Report No. UCLA-ENG-7653, University of California, Los Angeles, June 1976.

Lew, M., R. Chieruzzi, and K.W. Campbell, "Correlations of Seismic Velocity with Depth", Proceedings of the International Conference on Recent Advances in Geotechnical Earthquake Engineering and Soil Dynamics, St. Louis, Missouri, April 1981.

Richart, F.E., J.R. Hall, and R.D. Woods, Vibrations of Soils and Foundations. Prentice-Hall, Inc., Englewood Cliffs, New Jersey, 1970.

Shannon & Wilson, Inc., and Agbabian Associates, "Geotechnical and Strong Motion Data from U.S. Accelerograph Stations, Volume 1, Ferndale, Cholame, and El Centro, California," Report No. NUREG-0029 prepared for the U.S. Nuclear Regulatory Commission, 1975.

-oOo-

SUBJECT INDEX

Page number refers to first page of paper.

AUTHOR INDEX

Page number refers to first page of paper.